Organic Agriculture for Sustainable Livelihoods

This book provides a timely analysis and assessment of the potential of organic agriculture (OA) for rural development and the improvement of livelihoods. It focuses on smallholders in developing countries and in countries of economic transition, but there is also coverage of and comparisons with developed countries. It covers market-oriented approaches and challenges for OA as part of high-value chains and as an agro-ecologically based development for improving food security. It demonstrates the often unrecognized roles that organic farming can play in climate change, food security and sovereignty, carbon sequestration, cost internalizations, ecosystems services, human health and the restoration of degraded landscapes.

The chapters specifically provide readers with:

- an overview of the state of research on OA from socio-economic, environmental and agro-ecological perspectives;
- an analysis of the current and potential role of OA in improving the livelihoods of farmers, in sustainable value-chain development, and in the implementation of agro-ecological methods;
- proposed strategies for exploiting and improving the potential of OA and overcoming the constraints for further development; and
- a review of the strengths and weaknesses of OA in a sustainable development context.

Niels Halberg is Director of the International Centre for Research in Organic Food Systems (ICROFS), Tjele, Denmark. His publications include the co-edited book *Global Development of Organic Agriculture* (2006).

Adrian Muller has a dual appointment as Senior Researcher at the Forschungsinstitut für biologischen Landbau (FiBL/Research Institute of Organic Agriculture), Frick, Switzerland, and Professor of Environmental Policy and Economics (PEPE), at the Swiss Federal Institute of Technology ETH Zurich, Switzerland.

About CTA

The Technical Centre for Agricultural and Rural Cooperation (CTA) is a joint international institution of the African, Caribbean and Pacific (ACP) Group of States and the European Union (EU). Its mission is to advance food and nutritional security, increase prosperity and encourage sound natural resource management in ACP countries. It provides access to information and knowledge, facilitates policy dialogue and strengthens the capacity of agricultural and rural development institutions and communities.

CTA operates under the framework of the Cotonou Agreement and is funded by the EU.

For more information on CTA, visit www.cta.int

Organic Agriculture for Sustainable Livelihoods

Edited by Niels Halberg
and Adrian Muller

Routledge
Taylor & Francis Group

LONDON AND NEW YORK

from Routledge

First published 2013 by Routledge
2 Park Square, Milton Park, Abingdon, Oxon OX14 4RN

Simultaneously published in the USA and Canada by Routledge
711 Third Avenue, New York, NY 10017

Routledge is an imprint of the Taylor & Francis Group, an informa business

British Library Cataloguing in Publication Data
A catalogue record for this book is available from the British Library

Library of Congress Cataloging in Publication Data
Organic agriculture for sustainable livelihoods / edited by Niels Halberg
and Adrian Muller.
        p. cm.
    "Simultaneously published in the USA and Canada"—T.p. verso.
    Includes bibliographical references and index.
    1. Organic farming—Social aspects. 2. Organic farming—Economic aspects.
    3. Organic farming—Environmental aspects. 4. Sustainable agriculture—
    Social aspects. 5. Sustainable agriculture—Economic aspects. 6. Sustainable
    agriculture—Environmental aspects. 7. Sustainability. 8. Quality of life.
    9. Rural development. 10. Sustainable development. I. Halberg, Niels.
    II. Müller, Adrian F., 1971–
    S605.5.O635 2012
    631.5'84—dc23   2012003542

ISBN: 978–1–84971–295–8 (hbk)
ISBN: 978–1–84971–296–5 (pbk)
ISBN: 978–0–203–12848–0 (ebk)

Typeset in Bembo
by Keystroke, Station Road, Wolverhampton

MIX
Paper from
responsible sources
FSC
www.fsc.org   FSC® C004839

Printed and bound by CPI Group (UK) Ltd, Croydon, CR0 4YY

Contents

Figures

Case study 1

Case study 2

Case study 3

Case study 4

Tables

Contributors

Lucimar S. de Abreu, Researcher, Embrapa Meio Ambiente (Brazilian Agricultural Research Corporation, Environment Section), Rodovia SP 340 Km 127,5, Caixa Postal 69, 13820-000 Jaguariúna, SP, lucimar@cnpma.embrapa.br

Gustavo Fonseca de Almeida, Department of Agroecology – Farming Systems, Aarhus University, Blichers Allé 20, DK-8830 Tjele, Denmark, gustavo.dealmeida@agrsci.dk

Manuel Amador, Corporación Educativa para el Desarrollo Costarricense (CEDECO), Montelimar de Guadalupe, San José de Costa Rica, manuel@cedeco.or.cr

Lise Andreasen, International Coordinator, International Centre for Research in Organic Food Systems (ICROFS), Blichers Allé 20, P.O. Box 50, DK-8830 Tjele, Denmark, Lise.Andreasen@icrofs.org

Stéphane Bellon, Researcher at INRA SAD (French National Institute for Agricultural Research, Sciences for Action and Development Department), Ecodevelopment Unit, Site Agroparc, 84914 Avignon Cedex 9, France, bellon@avignon.inra.fr

Jonathan Castro, Corporación Educativa para el Desarrollo Costarricense (CEDECO), Montelimar de Guadalupe, San José de Costa Rica, jonathan@cedeco.or.cr

Aage Dissing, Fasanvej 11, 9460 Brovst, Denmark, fasanvej11@gmail.com

Inge Lis Dissing, Fasanvej 11, 9460 Brovst, Denmark, fasanvej11@gmail.com

Henrik Egelyng, Danish Institute for International Studies (DIIS), Strandgade 56, DK-1401 Copenhagen, Denmark, egelyng@mail.tele.dk

Bo van Elzakker, Director, Department of Tropical Agriculture, Agro–Eco Louis Bolk Institute, Hoofdstraat 24, 3972 LA Driebergen, The Netherlands, b.vanelzakker@louisbolk.org

Frank Eyhorn, Co-Team Leader, Rural Economy/Head, Organic & Fairtrade Competence Centre, HELVETAS Swiss Intercooperation, Weinbergstrasse 22a, P.O. Box 3130, CH-8021 Zurich, Switzerland, frank.eyhorn@helvetas.org

Elsio Antonio Pereira de Figueiredo, Empresa Brasileira de Pesquisa Agropecuária (Embrapa), Ministério da Agricultura, Pecuária e Abastecimento, BR 153, KM 110, Caixa Postal 21, Sistrito de Tamanduá, 89700-000 – Concórdia, SC, elsio@cnpsa.embrapa.br

Maria Fernanda Fonseca, PESAGRO-RIO/Estação Experimental de Nova Friburgo, Rua Euclides, Solon de Pontes, 30 – Centro – 28601-970 – Nova Friburgo – RJ, Brazil, mfernanda@pesagro.rj.gov.br

Francisca George, University of Agriculture, Abeokuta (UNAAB), P.M.B. 2240, Abeokuta, Nigeria, adebukolageorge@yahoo.com

Niels Halberg, Director, International Centre for Research in Organic Food Systems (ICROFS), Blichers Allé 20, P.O. Box 50, DK-8830 Tjele, Denmark, Niels.Halberg@icrofs.org

John E. Hermansen, Head of Research Unit, Department of Agroecology, Aarhus University, Blichers Allé 20, P.O. Box 50, DK-8830 Tjele, Denmark, john.hermansen@agrsci.dk

Paul Rye Kledal, Director, Institute of Global Food & Farming, Rymarksvej 89, 1 tv, DK-2900 Hellerup, Denmark, paul@igff.dk

Marie Trydeman Knudsen, Department of Agroecology, Aarhus University, Blichers Allé 20, P.O. Box 50, DK-8830 Tjele, Denmark, mariet.knudsen@agrsci.dk

Vibeke Langer, Associate Professor, University of Copenhagen, Department of Agriculture and Ecology/Crop Science, Højbakkegård Allé 13, DK-2630 Taastrup, Denmark, vl@life.ku.dk

Luping Li, Research Fellow, Center for Chinese Agricultural Policy, Institute of Geographical Sciences and Natural Resources Research, Chinese Academy of Sciences, No Jia11, Datun Road, Anwai, Chaoyang District, Beijing 100101, China, llp.ccap@igsnrr.ac.cn

Stewart Lockie, Professor, School of Sociology, Haydon-Allen Building 22, College of Arts and Social Sciences, The Australian National University, Canberra ACT 0200, Australia, stewart.lockie@anu.edu.au

Kristen Lyons, School of Social Science, University of Queensland, St Lucia, Brisbane QLD 407, Australia, Kristen.Lyons@uq.edu.au

Adrian Muller, Socio-Economics/Climate Change, Research Institute of Organic Agriculture (FiBL), Ackerstrasse, CH-5070 Frick, Switzerland; and Professor of Environmental Policy and Economics (PEPE), Department of Environmental Sciences, Institute for Environmental Decisions (IED), Swiss Federal Institute of Technology (ETH Zurich), Universitätsstrasse 22, 8092 Zürich, Switzerland, adrian.mueller@fibl.org

Jane Nalunga, Training Officer, National Organic Agricultural Movement of Uganda (NOGAMU), P.O. Box 70071, Kampala, Uganda, admin@nogamu.org.ug

Myles Oelofse, Post-doctoral Researcher, University of Copenhagen, Faculty of Life Sciences, Department of Agriculture & Ecology, Thorvaldsensvej 40, DK-1871 Frederiksberg C, Denmark, myles@life.ku.dk

Balgis Osman-Elasha, Climate Change Adaptation Expert, Compliance and Safeguards Division, ORQR3, African Development Bank, 13 Avenue de Ghana, BP 323, 1002 Tunis Belvedere, Tunis, Tunisia, balgis@yahoo.com

Gomathy Palaniappan, Lecturer, School of Agriculture and Food Sciences, University of Queensland, Gatton, Queensland, Australia 4343, g.palaniappan @uq.edu.au

P. Panneerselvam, Senior Scientist, Department of Biodiversity, M S Swaminathan Research Foundation, Taramani, Chennai 600113, India, panneerkvt@gmail.com

Yuhui Qiao, PhD, Associate Professor, Resource and Environment College, China Agricultural University Center of Resources Utilization; and Environmental Engineer (CRUEE), No. 2 Yuanmingyuan Xilu, Haidian District Beijing, China, qiaoyh@cau.edu.cn

Christian Schader, Socio-Economics, Research Institute of Organic Agriculture (FiBL), Ackerstrasse, CH-5070 Frick, Switzerland, christian.schader@fibl.org

Charles Ssekyewa, Uganda Martyrs University, Faculty of Agriculture, P.O. Box 5498, Kampala, Uganda, cssekyewa@umu.ac.ug

Thaddeo Tibasiima, Training Officer, Sustainable Agriculture Trainers Network (SATNET), P.O. Box 884, Fort Portal Town, Uganda, satnet@utlonline.co.ug

Sophia Twarog, Economic Affairs Officer, Trade, Environment and Development Branch, UNCTAD/DITC, E.8015, Palais des Nations, CH-1211 Geneva 10, Switzerland, Sophia.Twarog@unctad.org

Mette Vaarst, Department of Animal Science – Epidemiology and Management, Aarhus University, Blichers Allé 20, DK-8830 Tjele, Denmark, mette.vaarst@agrsci.dk

S. Vaheesan, Programme Officer, Helvetas Sri Lanka, Swiss Association for International Cooperation, 7 Glen Aber Place, Colombo 4, Sri Lanka

K. Zoysa, Tea Research Institute, Kandy, Sri Lanka

Editors' preface

There is a battle of discourse going on in the agriculture and food sectors over the definition and development pathways towards sustainable and sufficient food production. The challenges ahead include increasing food production while improving the sustainability of agriculture and reducing the negative effects on the environment. Put simply, one school seeks to further intensify agriculture based on external inputs while optimizing resource utilization in order to minimize pollution in a continuation of the previous five decades of agricultural development. Within this school, environmental considerations should be taken into account, but as a trade-off with the overarching goal of increasing production, and results are assessed in terms of productivity measures such as single crop yields per hectare and –subsequently – by inputs use and emissions per kg product. The other school of thought, called by some the food sufficiency narrative, acknowledges the need for intensification, but this should be based primarily on a better understanding and application of agro-ecological principles and local knowledge – and access to resources and land for a diversity of farmers. This school aims at creating synergies between agricultural production and preservation of the natural resource base such as soil quality and biodiversity. It also highlights the important interdependence between food consumption patterns and land use on local, regional and global scales.

Organic agriculture may be considered a key approach, but not the only one, within the sufficiency school. This book attempts to answer the question whether organic agriculture – and agro-ecology in a broader sense – can contribute to improving food security and the livelihoods of smallholder farmers while preserving our natural capital and enhancing other eco-system services at the same time.

The book follows the line from a previous multi-author volume *Global Development of Organic Agriculture: Challenges and Prospects* (Halberg et al. (eds), 2006) which focused on the globalization challenges for organic agriculture and discussed this in relation to political ecology, ecological justice, ecological economics, veterinary practices and food security. This new book takes a deeper look at the actual forms and results of organic agriculture from the perspective of farmers' livelihoods and in comparison with the ideals and principles behind organic agriculture. All chapters have been developed in a dialogue process between several authors, who often did not know each other beforehand. The draft chapters were also submitted to a peer-review process.

We hope the book will inspire the organic movements to continuous development in line with organic principles and with a focus on growth of the sector based on robust systems and at the same time securing integrity and trustworthiness vis-à-vis policy makers and consumers. We also hope that policy makers, researchers and other stakeholders outside the organic sector will be inspired to study and support the use of organic principles for agricultural development in broad, so that the principles may become a reference for sustainable agricultural development – under different titles such as climate-smart agriculture, ecological or eco-functional intensification.

First of all, we want to thank all the contributors to this book. It's their texts and their corresponding time and effort that makes it valuable. We would also like to thank the numerous colleagues who have inspired this discussion and with whom we have exchanged knowledge and experiences over the years, not least at the events organized by ISOFAR, the International Society of Organic Agriculture Research. Moreover, we acknowledge the many pioneers of organic agriculture – the farmers, advisors, non-governmental organizations (NGOs), companies, researchers, teachers and policy makers – who have contributed to the successful development of the organic sector – and still do so by discussing and testing the organic principles in theory and practice.

Niels Halberg   Adrian Muller
ICROFS   FiBL

1 Organic agriculture, livelihoods and development

Niels Halberg and Adrian Muller

Introduction

The present global food system is far from ideal and it is not sustainable, especially because of the lack of food security for large populations, the distribution and use patterns of food, and the externalities in terms of pollution and overuse and destruction of critical resources. But, there are many promising examples and initiatives of alternative development pathways which seek to combine improved food security and rural development with sustainable use of natural resources and enhancement of ecosystems services. This book takes a critical look at one group of such initiatives, the development and promotion of organic farming systems. In particular, it aims at investigating to what extent and under which conditions organic agriculture may contribute to environmental improvement, rural development and better livelihoods. In this, we consider organic agriculture to be a 'laboratory' for development of future sustainable food production.

The definition of organic agriculture and description of the principles and forms of this production form are given below, following a brief overview of the problems at stake and the interdependencies between food production, natural resource management and poor rural people's livelihoods. Then a short outline of the chapters and case studies in this book is given.

Challenges and options for the current food crisis – and the coming one

Even though, technically speaking, enough food is produced today for the present world population of 7 billion people to have a sufficient diet in terms of calories and protein (Halberg et al., 2006b), we live in a world with approximately 925 million food-insecure people (FAO, 2010), which comprises on average more than 15 per cent of the population in the developing countries. The majority of these hungry people live in poor, rural communities of South Asia and Africa south of the Sahara and are often directly involved in producing food, in terms of crops, livestock and fisheries (FAO, 2011a). Parts of the food-insecure population live in countries that on the overall scale are food sufficient and may even be net food exporters. Thus, India is technically self-sufficient in food and presently has significant stocks while there are still more than 200 million food-insecure people (FAO, 2010; see Chapter 2 for some more information on India). This demonstrates that hunger is mainly a question of poverty and lack of access to food and not necessarily a lack of global food production and supply.

The point of this is that improved food security for many millions of poor families in rural areas is mainly a question of improving food sufficiency by improving agriculture, natural resource management and market access, and reducing poverty. This is a multi-factor challenge, which cannot be solved by improved agricultural practices alone, but is linked with health, sanitation, education and institution building (FAO, 2010). However, there is a growing understanding that increased investments in agricultural development targeting the smallholder farmers in developing countries is an important element in improving food security in rural areas (FAO, 2010; Beddington et al., 2011; De Schutter and Vanloqueren, 2011). As discussed by Knudsen et al. (2006), the gap between the most and the least productive farming systems as measured by simple yields per hectare has increased by a factor of 20 over the last 50 years. This is mainly caused by differences in access to technology, knowledge and markets which favour large-scale, mechanized and high-input farming systems over smallholder farms. Farmers with less than 2 hectares of land constitute more than 90 per cent of farmers and cover some 60 per cent of the agricultural land globally. The potential for increasing their productivity is huge (De Schutter and Vanloqueren, 2011).

The present hunger and malnutrition problem is significant already – and by no means new – and it also has proven difficult to solve partly due to its complexity and a lack of sufficient political will. Unfortunately there are even more dire challenges for future global food security. With an estimated global human population of approximately 9.2 billion in 2050 and – more important – increased global demand for livestock products, it will be a challenge to provide sufficient food and biomass. There is a need for higher total food production per area unit, though the actual amounts needed depend on developments in diets, livestock feeding practices and food waste (Halberg et al., 2006a; Nellemann et al., 2009; Beddington et al., 2011; Freibauer et al., 2011).

The challenge is aggravated by the present use of natural resources in agriculture which risks impacting negatively on the options for improving food production in many areas. Thus, it is estimated that approximately 2 billion hectares of agricultural land have been given up because of erosion, salinization and compaction over the last 25 years. The mismanagement continues, leaving another 12 million hectares with degraded soils, which contributes to food insecurity due to yield reductions, reduced efficiency of input use and micro-nutrient deficiency (Lal, 2009; Nelleman et al., 2009; Beddington et al., 2011). According to Lal (2009), there is a need for a paradigm shift in land husbandry and for principles and practices for soil management, but with the adoption of proven management options global soil resources are adequate to meet the food and nutritional needs of both the present and future population. Known options for improved soil management and human nutrition include such techniques as mulching and recycling of organic residues; improving soil structure and quality; water conservation and water use efficiency; agro-forestry and mixed farming; diversified cropping systems including the use of indigenous foods and genetically modified organisms (GMOs) high in nutrients; no-till agriculture; use of micronutrient-rich fertilizers, nano-enhanced Zeolites; inoculating soils for improved biological nitrogen fixation; microbial processes to increase P-uptake (Okalebo et al., 2006; Lal, 2009). Most of these options – though not all – are interesting for and in line with organic agriculture.

Many forms of agriculture also affect biodiversity negatively, even though the ecosystem services provided by diversity in cultivated and non-cultivated areas are important for pollination and control of crop pests and diseases (Millennium Ecosystem Assessment, 2005; Perfecto et al., 2009). The current speed of species extinction is considered one of the most alarming signals of unsustainable human behaviour and agriculture is partly responsible (Rockström et al., 2009; Millennium Ecosystem Assessment, 2005). Preservation of biodiversity is often seen as conflicting with agricultural practices, which has led some authors to propose that there is a competition between improving food production and preserving biodiversity, 'land sparing vs. land sharing' (Phalan et al., 2011). However, others argue that reconciling the needs for biodiversity preservation and food production is an option for improved resilience and food security due to the interlinkages in ecosystems service functions (Perfecto et al., 2009; Brussard et al., 2010). The Millennium Ecosystem Assessment (2005) mentions both approaches as necessary for the long-term preservation of endangered species and ecosystems. The synthesis report states that:

> Effective response strategies include sustainable intensification, which minimizes the need for expanding total area for production, allowing more area for biodiversity conservation. Practices such as integrated pest management, some forms of organic farming and protection of . . . non-cultivated habitats within farms can provide synergistic relationships between agriculture, domestic biodiversity and wild biodiversity.

However, the report cautions against taking this as the only approach because there is too little evidence that this effectively secures species diversity sufficiently at regional levels.

Thus, there is a need to develop agricultural practices that create synergies with preservation and utilization of biodiversity, so-called functional biodiversity. There are good examples and evidence of the potential for creating synergy between food production and biodiversity by promoting farming systems that benefit from planned diversity in crops and non-cultivated areas in terms of reduction in pest problems and increased resilience to yield depression from pests and erratic rainfall (Jackson et al., 2007; Perfecto et al., 2009; Chappel and LaValle, 2011; FAO, 2011c; Kahn et al., 2011).

Water is expected to be an important scarcity in future agricultural production in many areas due to current overuse and pollution, climate change, low levels of soil organic matter resulting in low water-holding capacity and insufficient infrastructure for water harvesting and storage and for irrigation (Nelleman et al., 2009; UN-Water, 2007; Postel, 2011).

The above challenges will presumably be aggravated by the consequences of climate change, especially because increased temperatures will reduce yields in major cereal crops in many of the current 'bread basket' regions and because rainfall patterns will become more unpredictable. Moreover, due to an increased incident of high-intensity rain events there is a greater risk of surface erosion on soils which cannot percolate water sufficiently and this again increases the challenge of retaining water for crop growth (Clements et al., 2011; Beddington et al., 2011). Therefore, it is highly necessary to manage soils to have a good structure, including sufficient content of organic matter. Thus, with the increased impact of climate change on regional and local scales there is a need to develop adapted farming systems which are resilient to larger inter-annual variation in rainfall, with farmers who have capacity for continuously developing their practices as a response to changes in the environment (Beddington et al., 2011). In recent FAO terminology this is called 'climate smart agriculture': 'Climate smart or development smart agriculture is one that ensures that agriculture transcends the multiple issues with which it is currently associated – GHG emissions, loss of biodiversity, water misuse, soil and land degradation and socio-economic inequities which are compromising the world's capacity to feed its population' (Neely, 2011; FAO, 2011c).

There is an increased understanding that the challenges of producing enough food and biomass while preserving soil, water and biodiversity necessary for ecosystem services cannot be solved by prevalent types of conventional agriculture. In a review by 400 scientists and experts supported by the World Bank, the Food and Agriculture Organization (FAO) of the United Nations (UN), the United Nations Environment Programme (UNEP) and the International Assessment of Agricultural Knowledge, Science and Technology for Development (IAASTD) (McIntyre et al., 2009) it was concluded that '*business as usual is not an option*' because degradation of ecosystems already now limits or reverses productivity gains from high-input agriculture and because a huge number of smallholder farmers are left without proper agricultural technologies and extension services. The report states that '*a fundamental shift in AKST* (agricultural knowledge science and technology) is required to successfully meet development and sustainability goals'; that 'research, innovation and extension should account better for the complexity of agricultural systems within the diverse social and ecological contexts' and that an interdisciplinary and agroecosystems approach to knowledge production and sharing will be important for solving these needs. 'Advances in AKST can help create synergy among agricultural growth, rural equity and environmental sustainability. Integrated approaches to AKST can help agriculture adapt to water scarcity, provide global food security, maintain ecosystems and provide sustainable livelihoods for the rural poor' (McIntyre et al., 2009).

Likewise, in 2009 the Committee on Agriculture (COAG) of the FAO consisting of member country representatives 'endorsed the proposal that public and private investments be made in *agroecological research*, at both national and international levels' and the committee stressed that 'an ecosystem approach be adopted in agricultural management in order to achieve *sustainable agriculture*, including integrated pest management, organic agriculture and other traditional and indigenous coping strategies that promote agroecosystem diversification and soil carbon sequestration' (FAO, 2009).

The UN Special Rapporteur on the Right to Food states that under-investment in the agricultural sectors in many developing countries has limited the necessary uptake of agro-ecological methods, which are knowledge intensive, and that 'extension services that teach farmers – often women – about agro-ecological practices are particularly vital' (De Schutter and Vanloqueren, 2011).

Against this background, as described above, there is a need for the development and adoption of farming systems that seek to create synergy between food production and sustaining ecosystem services and are more resilient to climate change. The quest for such systems has many labels such as climate-smart agriculture, agro-ecology, organic farming, conservation agriculture and no–till farming. This book will focus on organic agriculture, informal and certified, and will analyse to what extent this would be a good bet for smallholder farmers in light of the challenges described.

Two forms of organic agriculture systems

For a large proportion of smallholder farmers an additional challenge is that they suffer from lack of market access and limited access to inputs and extension services on top of a declining soil fertility combined with changing rainfall patterns. There is a need to create innovative value-chain partnerships which may link such farmers better with markets while targeting the inherent risks both in the supply-and-demand system and the intermediary agents. Private-sector led initiatives may improve smallholder farmers' market access, improve capacity for intensification and thus give economic development.

As indicated in the IAASTD report (as mentioned in the previous section), improved access to and involvement in knowledge creation and adaptation of agro-ecological methods are a prerequisite for smallholder farmers to benefit from S&T. Producing certified organic products for high-value markets may be considered such a vehicle for providing smallholder farmers with access to knowledge and technology as part of connecting to high-value markets via companies and it is important to verify whether such capacity building is taking place in reality.

Organic food and fibre is one of the fastest-growing high-value market chains with huge potential for benefiting a large number of smallholder farmers and processing companies in Asia and Africa (EPOPA, 2008; Willer and Kilcher, 2009). Organic agriculture is spreading among farmers in large parts of the world: from poor smallholders being trained in agro-ecological methods by non-governmental organizations (NGOs), to family farmers entering into commercial high-value chains via engagement with companies seeking organic products, to large-scale producers converting to organics using their existing market channels. Besides the global market there is an increasing demand in regional metropoles, partly via upmarket supermarkets and the tourist industry. However, it is not clear to what extent market-oriented smallholder farmers will be competitive in these markets and some experiences show that farmers in certified production schemes may have limited knowledge regarding organic agricultural system development and agro-ecological practices.

According to Willer and Kilcher (2011) globally 37.2 million hectares of land in 160 countries was certified organic in 2009, which is 2 million more than the year before and more than three times the organic land certified in 1999. Europe is still the continent with the largest percentage of agricultural land being certified (1.9 per cent) and it saw an increase of almost a million hectares from 2008 to 2009 (12 per cent). However, in Africa the area rose 20 per cent to just over 1 million hectares and in Asia the certified organic land increased 230,000 hectares in the one year to approximately 3.6 million hectares (a 7 per cent increase). These numbers do not include the areas used for wild collection (e.g. berries, nuts, palmito, medicinal plants) and beekeeping for certified products, which in Africa alone is estimated to cover more than 16 million hectares. Thus, the area of certified organic land is on the rise in developing countries and the major part of this is due to conversion of smallholder farms.

The global number of certified organic farmers increased to 1.8 million in 2009, partly due to a large increase in organic farmers in India, where there are now some 677,000 producers; the largest number in any country. Second and third to India are Uganda with 188,000 and Mexico with 129,000 organic producers (Willer and Kilcher, 2011). Other countries with many certified organic farmers are Ethiopia and Tanzania (101,000 and 85,000 respectively). The development in Africa and India has been supported by NGOs funded by foreign aid donors. Moreover, national organic organizations have been established in East Africa and facilitated the formulation and adoption of a common East African standard for organic products in collaboration with governments and the UNCTAD (United Nations Conference on Trade and Development, Geneva). The major part of the organic products is exported to Europe, North America and Japan where the demand is still growing but increasingly there are market segments for certified organic products in the affluent middle classes in many countries including those in the BRIC (Brazil, Russia, India and China).

Besides the certified organic farms, a large number of farmers are practising informal forms of organic agriculture (OA), which loosely may be defined as farmers following the principles of OA but whose farms are not certified by an external certifier. As stated by the International Federation of Organic Agriculture Movements (IFOAM): 'Apart from third party certification there are other methods of organic quality assurance for the market place. These can be in the form of self-declarations or participatory guarantee systems. There are also situations where the relation between the consumer and the producers are strong enough to serve as a sufficient trust building mechanism, and no particular other verification is needed.' And further, from an official position by

IFOAM: 'Any system using the methods of Organic Agriculture and being based on the Principles of Organic Agriculture is regarded by IFOAM as "Organic Agriculture" and any farmer practicing such a system can be called an "organic farmer".'

Farmers may practise informal organic agriculture for different reasons, e.g. because they do not have access to a market paying price premiums for organic products and because they find that using organic and agro-ecological principles is a good solution for maintaining soil fertility and the health of crops and livestock and thus achieving food security without relying on cash-demanding inputs (Parrot et al., 2006). The extent of informal organic agriculture in terms of farmers and land area is difficult to estimate but may be many times larger than the certified organic area depending on the delimitation as discussed by Parrot et al. (2006). Many NGOs and development organizations introduce agro-ecological methods (see below) to smallholder farmers through for example farmer field school activities. A review of 286 organic and near-organic projects covered 12.5 million farmers on 40 million hectares (Pretty et al., 2006). A sub-sample of 57 projects in East Africa covered approximately 2 million hectares, which is twice the total certified land in Africa as recorded by Willer and Kilcher (2011).

Principles of organic agriculture

Some critics would say that poor uneducated farmers have always practised organic agriculture, which is equivalent to an unsustainable non-use of inputs, soil mining and unproductive traditional farming. But OA is *not* just the absence of chemical pesticides and fertilizer. The umbrella organization IFOAM, whose members are most of the national movements for organic agriculture and a number of companies and individuals, gives the following definition:

> Organic agriculture is a production system that sustains the health of soils, ecosystems and people. It relies on ecological processes, biodiversity and cycles adapted to local conditions, rather than the use of inputs with adverse effects. Organic agriculture combines tradition, innovation and science to benefit the shared environment and promote fair relationships and a good quality of life for all involved.
>
> (IFOAM, 2009)

This overall definition has been further specified by IFOAM in four guiding principles, Health, Ecology, Fairness and Care, which represent the basic aims which organic farming and food systems must strive to achieve (Luttikholt, 2007). As shown in Table 1.1, the Health and Ecology principles are about using agro-ecological methods such as recycling of nutrients and organic matter for improvement of soil health and enhancing functional diversity to minimize pests and diseases in crop production, and about proper management of the health and welfare of livestock. The principles of Fairness and Care are about

Table 1.1 The four guiding principles of organic agriculture as defined by IFOAM

	Health	Ecology	Fairness	Care
Principle	Organic agriculture should sustain and enhance the health of soil, plant, animal, human and planet as one and indivisible	Organic agriculture should be based on living ecological systems and cycles, work with them, emulate them and help sustain them	Organic agriculture should build on relationships that ensure fairness with regard to the common environment and life opportunities	Organic agriculture should be managed in a precautionary and responsible manner to protect the health and well-being of current and future generations and the environment
Keywords and concepts	Prevention, immunity, resilience, regeneration, interdependence between soil, plant and animal health	Recycling, efficient resource use, ecological balance, genetic and agricultural diversity, habitats, functional biodiversity	Socially and ecologically just use of natural resources and environment, allowing livestock to express natural behaviour	Technology assessment, and risk aversion, acknowledgement of limited understanding of ecosystems, respect for practical experience and indigenous knowledge

Source: IFOAM, 2009; for full explanations see Luttikholt, 2007.

the ethically and ecologically just use of natural resources and prudent use of new technology.

Similarly, the European Union has included objectives and principles of organic food and farming in a Council regulation (EC, 2007), which states:

Organic production shall pursue the following general objectives:

(a) establish a sustainable management system for agriculture that:
  (i) respects nature's systems and cycles and sustains and enhances the health of soil, water, plants and animals and the balance between them;
  (ii) contributes to a high level of biological diversity;
  (iii) makes responsible use of energy and the natural resources, such as water, soil, organic matter and air;
  (iv) respects high animal welfare standards and in particular meets animals' species-specific behavioural needs;

(b) aim at producing products of high quality.

These objectives have many similarities with the IFOAM definition of OA although the EC's focus is solely on biology, natural resources and animal welfare and does not include the ideas on care and ecological justice of IFOAM. This is also evident from the principles described in the EC regulation (2007, Article 4):

> Organic production shall be based on the following principles: a. the appropriate design and management of biological processes based on ecological systems using natural resources which are internal to the system; b. the restriction of the use of external inputs. Where external inputs are required or the appropriate management practices and methods referred to in paragraph (a) do not exist, these shall be limited to: (i) inputs from organic production; (ii) natural or naturally-derived substances; (iii) low solubility mineral fertilisers.

Under Article 5, a number of 'Specific principles applicable to farming' mention among others:

- 'the maintenance and enhancement of soil life and natural soil fertility, soil stability and soil biodiversity preventing and combating soil compaction and soil erosion, and the nourishing of plants primarily through the soil ecosystem';
- 'the recycling of wastes and by-products of plant and animal origin as input in plant and livestock production';
- 'the maintenance of plant health by preventative measures, such as the choice of appropriate species and varieties resistant to pests and diseases, appropriate crop rotations, mechanical and physical methods and the protection of natural enemies of pests';
- 'the maintenance of animal health by encouraging the natural immuno-logical defence of the animal, as well as the selection of appropriate breeds and husbandry practices'; and
- 'the application of animal husbandry practices, which enhance the immune system and strengthen the natural defence against diseases, in particular including regular exercise and access to open air areas and pastureland where appropriate'.

However, the EC regulation also includes a ban on GMOs and stipulates that practices and inputs should be 'based on risk assessment, and the use of pre-cautionary and preventive measures, when appropriate'. Moreover, the EC regulation mentions a goal of high product quality and includes a section on 'Specific principles applicable to processing of organic food' which mentions 'the processing of food with care, preferably with the use of biological, mechanical and physical methods'; 'the restriction of the use of food additives'; and 'the exclusion of substances and processing methods that might be mis-leading regarding the true nature of the product'.

Thus, the IFOAM and the European Commission have based the regulation of organic agriculture and the specific rules (e.g. of the allowed import of manure and bio-pesticides in crop production and the indoor and outdoor area per pig and hen) on a guiding set of ideas, in order to reflect that the specific rules are to be evaluated over time and tightened whenever possible due to developments in the sector and new knowledge becoming available. This dynamic approach to organic agriculture is an important premise for several chapters of this book. While the definitions and rules for organic agriculture historically have been developed in Europe and with parallels found in other industrialized countries (the USA, Canada, Japan and under the Codex Alimentarius Commission) this is increasingly matched by national and regional standards in, for example, Brazil and China and other developing countries. The regulations in Brazil and China are covered in Chapter 9 of this book. The definition of OA in the recently adopted East African Organic Products Standard (see Chapter 8) is a good example of the ambitious goals of organic agriculture and the aims of linking traditional knowledge and (agro-ecological) techniques with innovation:

> Organic agriculture is a holistic production management system, which promotes and enhances agroecosystem health, including bio-diversity, biological cycles and soil biological activity. It seeks to minimise the use of external inputs, avoiding the use of synthetic drugs, fertilisers and pesticides and aims at optimising the health and productivity of interdependent communities of soil life, plants, animals and people. It builds on East Africa's rich heritage of indigenous knowledge combined with modern science, technologies and practices.
>
> (East African Community, 2007)

Therefore, the concept of organic agriculture referred to in this book is one that is: ambitious beyond the abandonment of chemical fertilizers and pesticides; broader than the traditional focus on soil fertility; and dynamic in that its basic idea is as a development pathway, not a forever-fixed set of rules, which builds on a combination of modern science and farmers' own experiences including traditional knowledge. Several chapters will question to what extent the present forms of organic agriculture follow the principles and discuss what should be the next steps in order for the movement to be dynamic.

This is not a new discussion however. Since the 1990s it has been debated whether certified organic agriculture has been negatively influenced by its increased integration into large-scale and globalized market chains, symbolized by supermarkets and long-distance trade, and how this puts pressure on organic producers in terms of price, timely and continuous delivery systems, etc, which again favour large-scale operators, monoculture and specialization and intro-duces long-distance transport (Hall and Mogyorody, 2001; Rundgren, 2003; Milestad and Darnhofer, 2003; Raynolds, 2004; Knudsen et al., 2006; Darnhofer et al., 2009). Thus, there is evidence that this puts under pressure compliance with the organic principles, in terms of the health, ecology and fairness as discussed in Alrøe et al. (2006) and Halberg et al. (2006b). And so far, little in the regulation of certified organic agriculture and food secures the fairness of the chain operators towards smallholder farmers, farm workers and other weaker partners in the value chains.

Some authors also use the term 'organic agriculture' almost synonymously with 'agro-ecology', or say that organic farmers practise agro-ecological methods. However, this use of the term is controversial to other scientists and practitioners. Wezel et al. (2009) distinguish several different meanings of the term agro-ecology in the literature, some broader in their focus than others. Some authors use 'agro-ecology' to denote a scientific discipline where principles from ecology are used for the study of the relations between elements in a field (agronomy), in a farming system or – in a broader definition – in a food system including consumers. Thus, Altieri (2002) defined agro-ecology as 'the discipline that provides the basic ecological principles for how to study, design and manage agroecosystems that are both productive and natural resource conserving and that are also culturally sensitive, socially just and economically viable'. He continues:

> Agroecology goes beyond a one-dimensional view of agroecosystems – their genetics, agronomy, edaphology, etc. – to embrace an understanding of ecological and social levels of co-evolution, structure and function. Instead of focusing on one particular component of the agroecosystem, agroecology emphasizes the inter-relatedness of all agroecosystem components and the complex dynamics of ecological processes.

Besides the scientific interpretations of agro-ecology the term is increasingly used to denote practices in relation to environmentally friendly or sustainable agriculture, such as recycling of organic matter to improve soil and plant health and the enhancement of functional biodiversity for natural pest control. This is probably the reason for an overlap with terms such as organic agriculture or other types of alternative agriculture.

Moreover, in some countries, especially in Latin America, agro-ecology is also used to denote a social movement towards sustainable agriculture (Wezel et al., 2009). In this case, the term has a political dimension and includes a broader focus on a development strategy focused on environmentally friendly farming systems which enhance natural resources, biodiversity and the food sovereignty of the farm families as well as seeking empowerment of rural communities. Thus, Altieri (2002) also expects that the agro-ecological approach to natural resource management should simultaneously address poverty alleviation, food security, ecological management of resources and empowerment of rural communities. In this book, agro-ecology will be used in several different meanings in different chapters according to the authors. We find this a strength and follow Wezel et al.'s (2009) suggestion that the combination of scientific studies of agro-ecological systems and development of agro-ecological practices with social movements is a strength and in combination represents one of the main development pathways for defining future agricultural systems in a combination of proactive science and political movements.

The livelihood concept as inspiration and framework for the analyses

Throughout the book we make explicit and implicit reference to the concept of livelihood security and terms such as human, financial and natural capital. This builds on a conceptual framework defined in the realm of development policy and analysis as explained in the sustainable rural livelihood framework proposed by DFID (1999a, 1999b).

For our purposes in the context of rural livelihoods, poverty alleviation and development, a conceptual framework should address 'the diverse assets that rural people draw on in building livelihoods; the ways in which people are able to access, defend and sustain these assets; and the abilities of people to transform those assets into income, dignity, power and sustainability' (Bebbington, 1999). In addition, the framework should also reach across different scales. It should address 'the relationships between intrahousehold, household, regional and macro economies; and it should incorporate the relationships that households have with institutions and organizations that operate at wider scales' (Bebbington, 1999: 2029).

Livelihood security is framed via the capabilities, assets (stores, resources, claims and access) and activities required for a means of living. A livelihood is sustainable when people can cope with and recover from stress and shocks, maintain or enhance their capabilities and assets, and provide sustainable livelihood opportunities for the next generation (Chambers and Conway, 1992). Often, rural people assess livelihoods according to income criteria. However, cultural and social practices that accompany rural residence are equally important (Bebbington, 1999).

DFID (1999a, 1999b) proposed the sustainable rural livelihood framework based on the understanding that rural livelihoods need to account for people's access to five types of capital asset: financial, natural, human, physical and social. The framework attempts to conceptualize the main factors that affect poor people's livelihoods and the typical relationships between these factors. The basis for people's livelihoods is described in terms of these five types of capital assets shown as a pentagram at the left side of Figure 1.1. Human capital is the personal physical and mental capacity of the family; in our case a farmer including their skills and knowledge, whereas social capital describes the resources and capacities of the local society they belong to (e.g. self-help groups, village saving schemes, etc.). The physical and natural capital includes the farm resources in terms of land, livestock, buildings and equipment and perennial vegetation, soil fertility, non-cultivated areas for functional biodiversity, etc. Financial capital is the income and monetary savings of the family, which may be reinvested in productive activities, family welfare (schooling, medical services, sanitary equipment) or consumer goods.

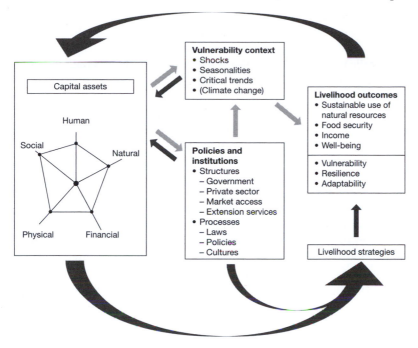

Figure 1.1 Sustainable livelihoods framework
Source: Revised after DFID, 1999a, 1999b.

Livelihood outcomes are primarily influenced by the capital assets and vice versa. The extent of people's access to capital assets is strongly influenced by their vulnerability context, and policies and institutions. Vulnerability context takes account of trends and changes (market, political, technological, globalization), shocks (epidemics, natural disasters), seasonality (prices, production, employment opportunities) and climate change. Access to capital assets is influenced by the prevailing social, institutional and political environment, which affects the ways in which people combine and use their assets to achieve livelihood outcomes. Livelihood strategies are combining the assets, taking into account the vulnerability context, supported or obstructed by policies, institutions and processes to achieve the livelihood outcomes. Vulnerability context, policies and institutions have a strong influence on people's access to capital assets and vice versa. For example, favourable policies, market access and extension services help the people to increase financial, human and physical capital. However, people with better social, human and physical capital influences the policies and market access, and people with strong financial and natural capital may have more resilience to shocks and climate change.

Eyhorn (2007) amended this framework by emphasizing the importance of an 'extended dimension of meaning' (Högger, 2000: 11), accounting for non-economic aspects such as culture, traditions, status considerations, conceptions of dignity and pride or personal ambitions. In his presentation of the sustainable livelihood framework, Eyhorn (2007) also broadens the focus by shifting from vulnerabilities that only affect livelihoods to both risks and opportunities. This makes explicit the potential for improvements – and not only aggravation – from any changes that affect the household. Moreover, we find that power and inequality aspects should be explicit elements of the conceptual framework. These factors can affect livelihoods in several ways: they play a role in the relations of an individual to other people and to the community; they play a role in the distribution of capital types, access to markets and other exchange contexts and thus affect the knowledge/activity and physical base and the socio-economic space. Power relations and inequalities have repercussions on the individual level, thus affecting the individual orientation, inner human space and emotional base. Finally, power relations and inequality can be decisive in the relation to institutions, organizations, etc. In this light, it is interesting to ask whether the organization of organic market chains and/or training in agro-ecological methods in itself may contribute to changed power relations and empowerment of smallholder farmers.

The sustainable livelihoods framework is thus a conceptual framework that helps to improve understanding of the livelihoods of rural poor people in developing countries. The sustainable rural livelihood framework and its amendments attempt to conceptualize the main factors that affect poor people's livelihoods and the typical relationships between these factors.

Purpose and outline of the book

On this background, the main purpose of this book is to scientifically investigate to what extent and under which conditions organic and/or agro-ecological agriculture may contribute to improved rural development and better livelihoods, both for people and for their environment.

This overall question may be formulated more specifically depending on perspective and interests and answering this will need a wide array of scientific disciplines. For example; a consumer in a Northern supermarket may wonder whether paying the price premium for certified organic coffee or bananas really improves the local environment in the country of origin and supports local development and poor farmers' livelihood and food security. The same question may be asked by NGOs aiming at improving poor people's livelihoods. Is organic agriculture really able to do the trick? There is presently some evidence that adoption of agro-ecological practices may increase the yields and food security for poor farming communities (Pretty et al., 2006) and that they reduce environmental degradation and this has been picked up by certain UN bodies including UNEP/UNCTAD (United Nations Conference on Trade and Development) (2008) and the UN Special Rapporteur on the Right to Food (De Schutter and Vanloqueren, 2011). However, more knowledge is needed on whether market-oriented, cash-crop focused organic agriculture improves smallholder farmers' food security and other aspects of livelihood.

These questions are treated in the first part of the book where Chapter 2 gives an overview of the literature studying the impact of organic agriculture adoption on food security. Chapter 3 presents evidence of the climate impact and other environmental assessments of organic products in countries with significant exports to Europe and North America at the farm level and along the food chain. Related to this, Chapter 4 presents knowledge on the extent to which the principles of health and ecology are backed by development and use of agro-ecological practices in smallholder organic farms in developing countries and discusses how these principles could be operationalized into a set of indicators describing the degree of using organic practices in farms. Following this, Chapter 5 explains the potential of OA for assisting farmers in adapting to climate change.

Chapter 6 discusses the degree of empowerment of smallholder farmers when entering global high-value food chains, based on different approaches of organic certification and the importance of power relations in successfully contributing to optimize the organic rules and regulations for local conditions.

Linked to this is the question under which circumstances organic agriculture may support smallholder farmers to increasingly influence their own market access. Chapter 7 presents research that illustrates the challenges for smallholder farmers in accessing high-value market chains such as organic products sold in supermarkets.

After this broad range of topics related to the scale of the farm and the value chains, the second part of the book looks at the societal scale and addresses general policy aspects and social movements. Given that policy makers and organizations decide to support OA development in a country or region there is a need for an institutional framework in terms of definitions and regulation of OA, accreditation and certification and support to research, development, adaptation of methods and dissemination. Chapter 8 gives an overall picture of what countries could or should do in order to support this development and Chapter 9 exemplifies this through a comparative analysis of the institutional frameworks for OA in two of the largest countries in terms of OA, China and Brazil. Chapter 10 describes the potential that 'social movements' may have in the development towards a more sustainable agriculture. The authors take the example of the agro-ecology movement in Latin America and show how agro-ecology is considered 'more organic than certified organics'. Finally, Chapter 11, addresses research needs in organic agriculture, focusing on Sub-Saharan Africa.

Throughout the book case studies supplement the text of the chapters with examples of organic agriculture. Some short cases have been inserted in smaller boxes within chapters to exemplify some of the points made by the authors. Other cases are presented in separate sections and have been commissioned to give in-depth information on case studies and projects, which complement the chapters. Thus, Case study 1 presents the work of NGOs using Farmer Field School (FFS) methods to engage poor farmers in western Uganda in the adaptation and use of agro-ecological methods. The authors describe how the focus changed from technical questions to social and economic ones and to the establishment of learning and savings groups in the villages. This led to a change in the set-up in the second project phase mainly reflected by the replacement of the FFS with so-called Family Learning Groups. The case represents a good example of the wider context of supporting resilient and adaptive rural communities based on an agro-ecology approach in the widest understanding and it is also a good example of improved livelihood and food security. It thus complements Chapters 2, 4, 6 and 10.

Case study 2 is a comparison of different types of orange producers in São Paulo state and exemplifies the points made in Chapter 4 that types of certified organic farms differ in how closely they follow the organic principles and that there is a need for indicators to assess these differences and facilitate comparisons and benchmarking between organic farms. Besides this the case is another example of the life cycle assessment of organic food chains demonstrated in Chapter 3.

Case study 3 presents an example of an NGO working with smallholder farmers to reduce their climate impact and, in a future step, adapt to climate change and develop a procedure allowing the farmers to benefit from this by a scheme under the voluntary carbon credits market. This case supplements the chapters on climate change adaptation and on environmental and climate change impact (Chapters 3 and 5).

Case study 4 shows results from a survey of organic tea producers in Sri Lanka, who use agro-ecological methods in a well-developed agro-forestry system. This case demonstrates a positive linkage between an agro-ecological approach developed with the assistance of a local NGO and certified organic production driven by a commercial company, which has the market link. This way it exemplifies points made in Chapters 4, 6 and 7. Moreover, the description of the significance of the FAIR Trade component of the tea scheme is complementary to the discussion of agroecology as a movement (in Chapter 10) and the limitations of organic certification schemes vis-à-vis this aspect.

From the definition of OA and the observation that it is not sufficient to abandon chemical inputs it follows that OA should be under continuous development and looking for new agroecological methods, improved animal welfare, product quality and securing improvement in human livelihoods. Moreover, farmers should be able to adapt their systems to changes in the economic and natural environment, and not least the influence of climate change. This requires ongoing research, development and adoption. One aspect of this has been coined by the term 'eco-functional intensification' in the European-based technology platform TP Organics: 'Eco-functional intensification is an approach which *aims to harness beneficial activities of the ecosystem to increase the productivity of agricultural systems. It is a new area of agricultural research, and represents an approach to global food security which aligns stability of supply with maintenance* of ecosystem services' (Padel et al., 2010). This approach is increasingly mirrored by other institutions which now talk about ecological intensification or 'sustainable crop production intensification' (FAO, 2011b). Thus, the former Director General of FAO, J. Diouf writes:

Sustainable intensification means a productive agriculture that conserves and enhances natural resources. It uses an ecosystem approach that draws on nature's contribution to crop growth – soil organic matter, water flow regulation, pollination and natural predation of pests – and applies appropriate external inputs at the right time, in the right amount . . . SCPI represents a major shift from the homogeneous model of crop production to knowledge-intensive, often location-specific, farming systems.

(FAO 2011b, Foreword)

Whether one foresees the need for chemical input or not, there seems to be increasing consensus that the starting point for intensification of poor smallholder farming in difficult environments is to develop, adapt and adopt agro-ecological practices. This is equivalent to the principles of organic agriculture as described above. Thus, there is a great need for a comprehensive strategic research agenda for organic agriculture in tropical and subtropical regions with smallholder farmers, not least in Sub-Saharan Africa. Chapter 11 gives a first attempt at highlighting research needs.

The combination of the need for further development of agro-ecological methods and other aspects of the organic principles in research and in practice points to one significant role of organic agriculture in relation to the future of food production. This mirrors important conclusions in the IAASTD report (McIntyre et al., 2009), by FAO and by the UN Special Rapporteur on the Right to Food as discussed in this chapter. One way to look at organic agriculture and agro-ecology is, therefore, as a laboratory for development of the future sustainable food production, and certification is a means of getting consumers onboard and contributing to this by paying price premiums. Thus, the chapters in this book aim at scrutinizing this picture of organic agriculture from different perspectives and test whether the vision and tangible results are still in sight.

References

Alrøe, H.F., Byrne, J. and Glover, L. (2006) 'Organic agriculture and ecological justice: ethics and practice', in N. Halberg, H.F. Alrøe, M.T. Knudsen and E.S. Kristensen (eds), *Global Development of Organic Agriculture. Challenges and Prospects*, CABI Publishing, Wallingford

Altieri, M.A. (2002) 'Agroecology: the science of natural resource management for poor farmers in marginal environments', *Agriculture, Ecosystems and Environment*, 93, pp. 1–24

Bebbington, A. (1999) 'Capitals and capabilities: a framework for analyzing peasant viability, rural livelihoods and poverty', *World Development*, 27, 12, pp. 2021–2044

Beddington, J., Asaduzzaman, M., Fernandez, A., Clark, M., Guillou, M., Jahn, M., et al. (2011) 'Achieving food security in the face of climate change', Summary for policy makers from the Commission on Sustainable Agriculture and Climate Change, CGIAR Research Program on Climate Change, Agriculture and Food Security (CCAFS), Copenhagen, Denmark, www.ccafs.cgiar.org/commission (accessed December 4, 2011)

Brussaard, L., Caron, P., Campbell, B., Lipper, L., Mainka, S., Rabbinge, R., et al. (2010) 'Reconciling biodiversity conservation and food security: scientific challenges for a new agriculture', *Current Opinion in Environmental Sustainability 2010*, 2, 34–42

Chambers, R. and Conway, G. (1992) 'Substantial rural livelihoods: practical concepts for the 21st century', Discussion Paper 296, Institute for Development Studies, University of Sussex, Brighton

Chappel, M.J. and LaValle, L. (2011) 'Food security and biodiversity: can we have both? An agroecological analysis', *Agriculture and Human Values*, 28, 3, pp. 3–26

Clements, R., Haggar, J., Quezada, A. and Torres, J. (2011) *Technologies for Climate Change Adaptation – Agriculture Sector*, X. Zhu (ed.) UNEP Risø Centre, Roskilde

Darnhofer, I., Lindenthal, T., Bartel-Kratochvil, R. and Zollitsch, W. (2010) 'Conventionalisation of organic farming practices: from structural criteria towards an assessment based on organic principles, A review', *Agronomy for Sustainable Development*, 30, 1, pp. 67–81

De Schutter, O. and Vanloqueren, G. (2011) 'The new Green Revolution: how twenty-first-century science can feed the world', *Solutions for a Sustainable and Desirable Future*, 2, 4, pp. 33–44, www.thesolutionsjournal.com/node/971 (accessed December 4, 2011)

DFID (1999a) 'Sustainable Livelihoods and Poverty Elimination', Department for International Development, London

DFID (1999b) 'Sustainable Livelihoods Guidance Sheets', Department for International Development, London, www.livelihoods.org/info/info_guidancesheets.html (accessed December 4, 2011)

East African Community (EAC) (2007) 'East African organic products standard', EAS 456:2007

EC (2007) 'Council Regulation (EC) No 834/2007 of 28 June 2007 on organic production and labelling of organic products and repealing Regulation (EEC) No 2092/91' Official Journal of the European Union, L 189/1, pp.1–23

EPOPA (2008) 'Organic exports – a way to a better life? Export promotion of organic products from Africa', Export Promotion of Organic Products from Africa (EPOPA), p. 105

Eyhorn, F. (2007) 'Organic farming for sustainable livelihoods in developing countries: the case of cotton in India', Inaugural dissertation, Philosophisch-naturwissenschaftlichen Fakultät der Universität Bern

FAO (2009) 'Report of the twenty-first session of the Committee on Agriculture (COAG)', ftp://ftp.fao.org/docrep/fao/meeting/016/k4952e.pdf (approached Nov. 28, 2011) (accessed December 4, 2011)

FAO (2010) 'The State of Food Insecurity in the World 2010, Addressing food insecurity in protracted crises', Food and Agriculture Organization of the United Nations, Rome, www.fao.org/docrep/013/i1683e/i1683e00.htm (accessed December 4, 2011)

FAO (2011a) 'The State of Food Insecurity in the World 2011, How does international price volatility affect domestic economies and food security', Food and Agriculture Organization of the United Nations, Rome, www.fao.org/docrep/014/i2330e/i2330e00.htm (accessed December 4, 2011)

FAO (2011b) *Save and grow: a policymaker's guide to the sustainable intensification of smallholder crop production,* Food and Agriculture Organization of the United Nations, Rome

FAO (2011c) *Climate change and food systems resilience in Sub-Saharan Africa,* L.L. Ching, S. Edwards and N.E.-H. Scialabba (eds), Food and Agriculture Organization of the United Nations, Rome

Freibauer, A., Mathijs, E., Brunori, G., Damianova, Z., Faroult, E., Girona i Gomis, J. et al. (2011) 'Sustainable food consumption and production in a resource-constrained world', European Commission – Standing Committee on Agricultural Research (SCAR), The 3rd SCAR Foresight Exercise, p. 150, http://ec.europa.eu/research/agriculture/scar/pdf/scar_feg3_final_report_01_02_2011.pdf (accessed December 4, 2011)

Halberg, N., Alrøe, H.F. and Kristensen, E.S. (2006a) 'Synthesis: prospects for organic agriculture in a global context', in N. Halberg, H.F. Alrøe, M.T. Knudsen and E.S. Kristensen (eds), *Global Development of Organic Agriculture: Challenges and Prospects*, CABI Publishing, Wallingford, pp. 344–364

Halberg, N., Sulser, T.B., Høgh-Jensen, H., Rosegrant, M.W. and Knudsen, M.T. (2006b) 'The impact of organic farming on food security in a regional and global perspective', in N. Halberg, H.F. Alrøe, M.T. Knudsen and E.S. Kristensen (eds), *Global Development of Organic Agriculture. Challenges and Prospects*, CABI Publishing, Wallingford, pp. 277–316

Hall, A. and Mogyorody, V. (2001) 'Organic farmers in Ontario: an examination of the conventionalization argument', *Sociologia Ruralis*, 41, 4, pp. 399–422

Högger, R. (2000) 'Understanding livelihood systems as complex wholes', in Högger, R. and Baumgartner, R., (eds) *In search of sustainable livelihood systems*, SAGE, New Delhi.

IFOAM (2009) 'Definition of Organic Agriculture, www.ifoam.org/growing_organic/definitions/doa/index.html (accessed December 4, 2011)

Jackson, L.E., Pascual, U. and Hodgkin, T. (2007) 'Utilizing and conserving agrobiodiversity in agricultural landscapes', *Agriculture, Ecosystems and Environment*, 121, pp. 196–210

Kahn, Z., Midega, C., Pittchar, J., Pickett, J. and Bruce, T. (2011) 'Push-pull technology: a conservation agriculture approach for integrated management of insect pests, weeds and soil health in Africa. UK government's Foresight Food and Farming Futures project', *International Journal of Agricultural Sustainability*, 9, 1, pp. 162–170

Knudsen, M.T., Halberg, N., Olesen, J.E., Byrne, J., Venkatesh, I. and Toly, N. (2006) ' Global trends in agriculture and food systems', in N. Halberg, H.F. Alrøe, M.T. Knudsen and E.S. Kristensen (eds), *Global Development of Organic Agriculture. Challenges and Prospects*, CABI Publishing, Wallingford

Lal, R. (2009) 'Soil degradation as a reason for inadequate human nutrition', *Food Security*, 1, pp. 45–57, DOI: 10.1007/s12571-009-0009-z

Luttikholt, L.W.M. (2007) 'Principles of organic agriculture as formulated by the International Federation of Organic Agriculture Movements', NJAS 54–4, *Wageningen Journal of Life Sciences*

McIntyre, B., Herren, H.R., Wakhungu, J. and Watson, R.T. (eds) (2009) 'Agriculture at a crossroads', Synthesis Report and Global Summary for Decision-makers, IAASTD, www.agassessment.org/reports/IAASTD/EN/Agriculture%20at%20a%20Crossroads_Synthesis%20Report%20(English).pdf (accessed December 4, 2011)

Milestad, R. and Darnhofer, I. (2003) 'Building farm resilience: the prospects and challenges of organic farming', *Journal of Sustainable Agriculture*, 22, 3, pp. 81–97

Millennium Ecosystem Assessment (2005) 'Ecosystems and human wellbeing: biodiversity', Synthesis, World Resources Institute, Washington, www.maweb.org/en/Synthesis.aspx (accessed December 4, 2011)

Neely, C. (2011) 'Increasing agriculture's climate smartness', Summary Notes from the Workshop on Agriculture Development and Climate Smart Agriculture in Developing Countries, Copenhagen, February, funded by the FAO and the

European Union and organized by Aarhus University, ICROFS, CCAFS, the University of Copenhagen and the FAO

Nelleman, C., MacDevette, M., Manders, T., Eickhout, B., Sivhus, B., Prins, A.G., et al. (eds) (2009) 'The environmental food crisis – the environment's role in averting future food crises', A UNEP rapid response assessment, United Nations Environment Programme, GRID-Arendal, www.grida.no (accessed December 4, 2011)

Okalebo, J.R., Othieno, C.O., Woomer, P.L., Karanja, N.K., Semoka, J.R.M., Bekunda, M.A., et al. (2006) 'Available technologies to replenish soil fertility in East Africa', *Nutrient Cycling in Agrosystems*, 76, pp. 153–170

Padel, S., Pearce, B., Schluter, M. Schmid, O., Cuoco, E., Willer, H. et al. (2010) 'Research themes to address food and agriculture challenges in the 8th EC', in *Implementation Action Plan for Organic Food and Farming Research,* TP Organics, www.tporganics.eu/upload/TPOrganics_ImplementationActionPlan.pdf (accessed December 4, 2011)

Parrot, N., Olesen, J.E. and Høgh-Jensen, H. (2006) 'Certified and non-certified organic farming in the developing world', in N. Halberg, H.F. Alrøe, M.T. Knudsen and E.S. Kristensen (eds), *Global Development of Organic Agriculture. Challenges and Prospects*, CABI Publishing, Wallingford

Perfecto, I, Vandermeer, J. and Wright, A. (2009) *Nature's Matrix. Linking Agriculture, Conservation and Food Sovereignty*, Earthscan, London

Phalan, B., Onial, M., Balmford, A. and Green, E. (2011) 'Reconciling food production and biodiversity conservation: land sharing and land sparing compared', *Science*, 333, 1289, doi: 10.1126/science.1208742

Postel, S.L. (2011) 'Getting more crop per drop', in L. Starke (ed.) *2011 State of the World, Innovations that Nourish the Planet,* Earthscan, London

Pretty, J.N., Noble, A.D., Bossio, D., Dixon, J., Hine, R.E., Penning de Vries, F.W.T. and Morison, J.I.L. (2006) 'Resource-conserving agriculture increases yields in developing countries', *Environmental Science & Technology*, 40, 4, pp. 1114–1119

Raynolds, L.T. (2004) 'The globalization of organic agro-food networks', *World Development*, 32, 5, pp. 725–743

Rockström, J., Steffen, W., Noone, K., Persson, Å, Chapin, F.S., Lambin, E.F. et al. (2009) 'A safe operating space for humanity', *Nature*, 461, pp. 472–475

Rundgren, G. (2003) 'Trade and marketing, no simple solution', *Ecology and Farming*, 34, pp. 6–8

UN-Water (2007) 'Coping with water scarcity. Challenge of the twenty-first century', Food and Agriculture Organization (FAO), www.un.org/waterforlifedecade/scarcity.shtml (accessed December 4, 2011)

UNEP/UNCTAD (2008) 'Organic agriculture and food security in Africa', Capacity Building Task Force on Trade, Environment and Development (CBTF). United Nations Conference on Trade and Development. United Nations Environment Programme, www.unctad.org/en/docs/ditcted200715_en.pdf (accessed December 4, 2011)

Wezel, A., Bellon, S., Doré, T., Francis, C., Vallod, D. and David, C. (2009) 'Agroecology as a science, a movement and a practice. A review', *Agronomy for Sustainable Development*, 29, 4, pp. 503–515

Willer, H. and Kilcher, L. (eds) (2009) *The world of organic agriculture. Statistics and emerging trends 2009*, IFOAM and FiBL, Bonn and Frick, ITC, Geneva, p. 299

Willer, H. and Kilcher, L. (eds) (2011) *The world of organic agriculture. Statistics and emerging trends 2011*, FiBL-IFOAM report, IFOAM and FiBL, Bonn and Frick, p. 286

2 Consequences of organic agriculture for smallholder farmers' livelihood and food security

P. Panneerselvam, Niels Halberg and Stewart Lockie

Introduction

While modern agricultural technology was developed with the noble intention of feeding the ever growing world population and has proven efficient in some parts of the world, until now it has failed – at the same time – to provide food security in many parts of the world. Modern agricultural methods have brought spectacular increases in productivity over the past 40 years; there has been remarkable growth in agricultural production with per capita world food production growing by 17 per cent and aggregate world food production growing by 145 per cent (FAO, 2005). On the other hand, more than 1 billion people (around one-sixth of the world population) are undernourished, 98 per cent of them from developing countries (FAO, 2010). Schmidhuber and Tubiello (2007) stated that the crucial issue of food security is not whether food is 'available' but whether the monetary and non-monetary resources at the disposal of the population are sufficient to allow everyone access to adequate quantities of food. 'National sufficiency is neither necessary nor sufficient to guarantee food security at individual level. For example, Hong Kong and Singapore are not self sufficient (agriculture is non-existent) but their populations are food secure' (Schmidhuber and Tubiello, 2007: 19703), whereas India as a country is self-sufficient in food production but home to 231 million food-insecure people. Evans (2009) argued that the benefits of modern technology have been unequally distributed, with large farms reaping the major benefits and leaving the majority of small farms untouched. This highlights the fact that modern technologies could not entirely address the issues faced by small farms in developing countries.

Small farms constitute 90 per cent of the world's farms. They provide employment to 1.3 billion people and dominate agriculture in developing countries. About two-thirds of the 3 billion rural people in the world live off the income generated by farmers managing some 500 million small farms of less than 2 ha each (FAO, 2008a). For smallholders, agriculture is not only a source of food but also a source of income to access other livelihoods. The main limitation of smallholder farming in developing countries is poverty and social exclusion along with lack of access to land and other productive resources. Hence, 75 per cent of the world's poor live in rural areas and 90 per cent of them are smallholders (FAO, 2008a). Furthermore, the twenty-first century adds new challenges to the resource-poor smallholders as a consequence of global climate change. Climate change will have more negative impact on resource poor smallholders living in semi-arid and sub-humid areas of developing countries meaning that the poorest regions with the highest level of undernourishment will be exposed to the highest degree of instability of food production due to climate change (Schmidhuber and Tubiello, 2007).

Lack of infrastructure, transport and market information, and poor farmer organization for bulk marketing are the main limiting factors in improving profitable marketing of surplus agricultural production and thereby increasing income in developing countries, particularly in Africa (Bolwig and Odeke, 2007; Warning and Key, 2002). Small farms are often limited by low risk-bearing capacity due to lack of access to crop insurance or other risk-reducing instruments to deal with production variability caused by climatic hazards. Another major concern for the majority of smallholders is lack of access to public credit which is why they often borrow from private money lenders at exorbitant interest rates (DES, 2009). Moreover, small farmers are often resource-poor and unable to buy expensive inputs without risk of indebtedness. The increase of input prices further worsens the situation, for example between 2006 and 2008, the rate of increase in the price of fertilizer was higher than the increase of food prices (Dorward and Poulton, 2008). Therefore, small-holders are in need of low-cost technology to reduce their dependency on high-input prices and reduce the vulnerability of production systems. Organic farming in developing countries is considered to reduce production cost by efficient use of farm resources and has the potential to improve the smallholders' livelihoods. Hence, the objective of this chapter is to discuss the potential of organic agriculture to improve smallholders' livelihoods in developing countries.

Different forms of practising organic farming

Organic agriculture includes both certified and non-certified food systems. Certified systems must be managed according to organic standards and principles and are subject to regular independent inspections. Farming systems that actively follow organic principles are considered organic, however, even if the agro-ecosystem or the farm is not formally certified as organic (Scialabba, 2007). IFOAM (2004) uses the term 'informal' or 'non-certified' organic agriculture for agricultural systems that meet the principles of organic agriculture but which are not certified. In other words non-certified organic agriculture refers to organic agricultural practices by *intent* and not by *default*. Informal organic farming is often promoted by NGOs and other agencies in developing countries. The produce of these systems is usually consumed by households or sold locally (e.g. urban and village markets) with little or no price premium.

A growing trend in developing countries is for certified organic farming to take place under contract farming arrangements (Bolwig et al., 2009; EPOPA, 2008). Typically, these oblige a firm to supply inputs, credits and information for smallholders, in exchange for a marketing agreement that fixes a price for the product and binds the farmer to follow a particular production method or input regimen (Warning and Key, 2002; Panneerselvam, 2011). Contract farming provides input credit directly to the farmers or farmers' associations with reimbursement at the time of product sale (Warning and Key, 2002). Often this is promoted as a means to improve livelihood/food security for smallholding farmers. On the other hand, contract farming has been criticized as a tool for agro-industrial firms to exploit an unequal power relationship with growers (Key and Runsten, 1999). However, contract farming is recommended as the alternative way to make small farming more productive as the services and inputs provided by the contract firms cannot be provided effectively by other agencies, and taking the social collateral in determining participation in income-generating activities into account, then poorer households are more able to participate and thereby reduce poverty and income inequality.

Livelihood outcomes and food security dimensions

Livelihood security is defined as the capabilities, assets (stores, resources, claims and access) and activities required for a means of living. A livelihood is sustainable when people can cope with and recover from stress and shocks, maintain or enhance their capabilities and assets, and provide sustainable livelihood opportunities for the next generation (Chambers and Conway, 1992). Often, rural people assess livelihoods according to income criteria. However, cultural and social practices that accompany rural residence are equally important (Bebbington, 1999). People require five types of asset: financial, natural, human, physical and social to achieve positive livelihood outcomes (DFID 1999a, 1999b).

One of the primary livelihood objectives for poor families is food security as described in the sustainable livelihood framework (Chapter 1). Food security exists when all people, at all times, have physical and economic access to sufficient safe and nutritious food that meets their dietary needs and food preferences for an active and healthy life (World Food Summit, 1996). This definition comprises four dimensions: food availability, access, utilization and stability. Household food security is when sufficient quantities of food are available for a family, either through household production (food availability), or because the family has adequate resources to buy food (food access), and when available and accessible food is properly used for a balanced diet for all household members (food utilization), and when households have access to adequate food at all times (food stability). All four dimensions must be fulfilled simultaneously to achieve the objectives of food security (FAO, 2008b).

In this chapter we address how organic agriculture influences income (financial capital), food security and other livelihood aspects including human capital.

The influence of OA on natural capital is addressed in Chapter 3 and social capital is addressed in Chapters 6 and 10.

Influence of organic agriculture on smallholders' income and food security

The FAO conference on organic agriculture and food security, May 2007, concluded that organic agriculture has the potential to improve food security in developing countries, particularly among small farmers who are often among the poorest and least food secure (Scialabba, 2007). Organic agriculture may improve local food security through the production of a diverse range of products at reduced input cost (IFAD, 2005). However, the relationships between organic production, poverty alleviation and food security may vary between certified and informal approaches to organic production and marketing. Halberg et al. (2009) suggest that certified organic production focused on cash cropping has the potential to improve food security primarily by increasing household income and purchasing power (see Table 2.1). Informal diversified organic food

Table 2.1 Possible contribution of organic agriculture on four dimensions of food security in relation to smallholders

Food security dimension	Certified organic agriculture	Informal (non-certified) organic agriculture
Physical AVAILABILITY of food (level of food production)	Focus on cash crops, moderate change in management and food crop yields	Focus on food crops, changed management, intensified land use, yield increases
Economic and physical ACCESS to food (focused on incomes, expenditure, markets and prices)	Increased household income, reduced production cost, improved market access with price premium	Increased income through reduced production cost and improved local market access with little or no price premium
Food UTILIZATION (diversity of the diet, nutrients in the food, health)	Diversified food purchases, inclusion of protein-rich legumes in the cropping pattern leads to increased diversity of diet, improving health by avoiding chemicals	Diversified food production, inclusion of legumes in the intercropping, avoiding chemicals, improving health by avoiding chemicals
STABILITY of the other three dimensions over time	Higher income and reduced debts lead to capital building and secure purchasing power. Increased resilience towards economic shocks in the family. But dependent on export markets and demand in the industrialized world	Diversity of crops and improved soil fertility lead to farm resilience towards pests and climate change Little capital building so still economically vulnerable

Source: After Halberg et al., 2009.

crop production for local markets, by contrast, promises to increase food security through increased local food production and sales and sometimes through increased income from local market sales. We might add that informal organic production is more likely to be associated with enhanced and/or lower cost production for subsistence purposes and non-commoditized exchange (i.e. gift economies, land rent, etc.). Importantly, however, these relationships are mediated by other livelihood strategies and may, in fact, interact (see Box 2.1). The rest of this section will elaborate on the contribution of certified and informal organic agriculture to the four dimensions of food security.

Box 2.1 Market and subsistence rice production in Negros Occidental, the Philippines

Some 347 farming households were surveyed in the uplands of Negros Occidental in 2007/08 on livelihood strategies and outcomes (Lockie and Carpenter, 2010). Households were drawn from areas that had been targeted by NGOs for assistance to convert to organic production. Major crops were rice, maize, fruit, vegetables and coffee. Households reported mean annual net incomes only marginally above the rural poverty line and most nominated a 2–4-month window every year of food insecurity. Organic and conventional rice farmers displayed similar levels of productivity (2.5 and 2.6 tonnes/ha/y respectively). However, organic producers achieved net returns 30 per cent higher on rice sold, due largely to a 49 per cent reduction in input costs (including labour). Where organic farmers were able to access a modest premium (approximately 10 per cent on unmilled rice) through NGOs coordinating domestic organic value chains, this improvement in net return increased to 50 per cent over conventional production. Interestingly, while the potential to access such premium markets was not a major motivating factor for most organic farmers (more important were opportunities to reduce costs and exposure to agrichemicals), these farmers sold a significantly higher proportion of their rice crop than did conventional farmers, keeping 47 per cent for self-consumption compared with 66 per cent among conventional farm households. Reflecting this, rice contributed a substantially higher share of total net household income on organic farms: 23 per cent as compared with 9 per cent. At face value, this suggests organic certification and markets have encouraged a shift away from subsistence and towards cash crop production of rice; the food security implication of which would be improved household purchasing power offset by declining local food availability. It is critical, however, to note: (1) that yield performance among both organic and conventional farmers varied dramatically; and (2) that very strong correlations were evident between the amount of rice households kept for self-consumption and the total amount grown, the amount sold, the land resources devoted to rice, the investment made in inputs (including labour and transport) and productivity per ha. In other words, those households making most use of rice as a subsistence crop were not those that kept a disproportionately large share of the harvest, but those that committed the most resources to its production and demonstrated the most capacity to manage these resources. This pattern was repeated with other staple crops such as maize. Organic farming did not therefore encourage a shift away from subsistence production of staple crops, but a shift away from livelihood strategies based on low-risk, low-productivity farming supplemented with wage labour.

Food availability

The capability of organic agriculture to produce sufficient amounts of food (food availability dimension) depends on physical productivity (yield) of the total sum of crops harvested. In other words, food availability depends on total amount of food harvested in a farm, which depends on the yield of each crop and the intensity of land use, i.e. intercropping in time and space. We have calculated and averaged the relative yields of cereals, pulses, oilseeds and other crops between organic and conventional systems from recently published papers specific to developing countries (Figure 2.1). Also, we have calculated the relative yield of individual crops between organic and conventional production (Figures 2.2 and 2.3). The relative yield is the ratio of organic to non-organic yield reported by the studies. For example, a relative yield of 0.80 means that organic yields were reported to be 80 per cent of conventional yields obtained for the same crop in a given area. The studies included mostly originated from Asia (India, Bangladesh, China and Sri Lanka), Brazil, Argentina and Kenya. The results show that the relative yields of fruits, vegetables and potato were higher in organic farming while the relative yields of cereals and oilseeds were lower

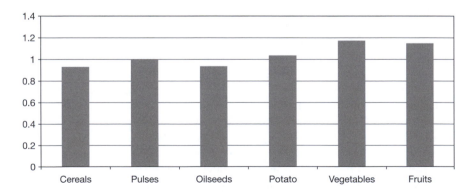

Figure 2.1 The relative yield of organic to conventional

(Figure 2.1). The relative yields of pulses were similar in organic and conventional farming.

The relative yields in organic farming are directly related to the intensity of prevailing farming systems and the levels of external inputs used before conversion (Halberg et al., 2006). Converting high-input farms generally results in lower relative yields while converting low-input systems often results in higher yields (Halberg et al., 2006; Badgley et al., 2007; Pretty et al., 2003). This is not only the case for comparison between regions, but also between crops within a region and for individual crops over time. Figures 2.1, 2.2 and 2.3 show that converting to organic farming resulted into yield reductions in wheat (India), oranges (Brazil), ginger (China) and peanut (India). In contrast, organic yields were higher in maize (Kenya), coffee (Mexico), banana (India), sugarcane and Beans (India), cabbage, potato, cauliflower and carrot (India). The relative yield of wheat was low in organic compared to other crops wherever it was grown. The wheat yield reduction in organic varied from 6 per cent (Eyhorn et al., 2007) to 40 per cent (Panneerselvam et al., 2011a) in India. Similarly, in Europe, the relative yield of wheat is often lower (55–80 per cent of conventional) compared with other crops partly due to the high nitrogen demands early in the growth season (Halberg and Kristensen, 1997).

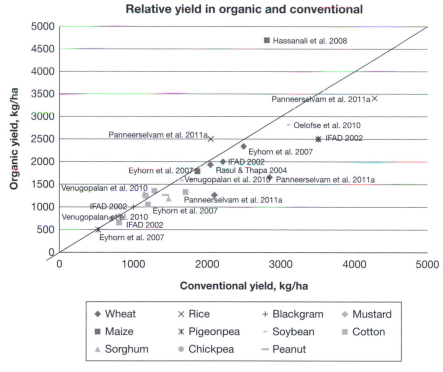

Figure 2.2 Relative yield of organic to conventional for cereals, oilseeds and other crops

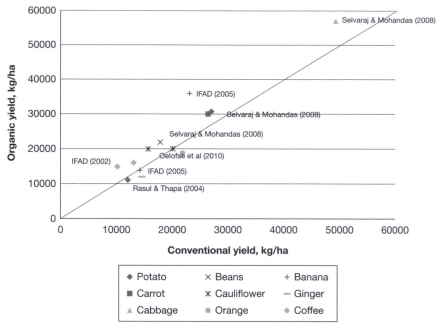

Figure 2.3 Relative yield of organic to conventional for vegetables, fruits and coffee

---

Box 2.2 bioRe India organic cotton project in Madhya Pradesh, India

bioRe India (www.bioreindia.com) is a private agribusiness company, promoting organic cotton production in Madhya Pradesh for export to Europe. The company started as a non-commercial experiment to help cotton farmers reduce debt and secure more sustainable livelihoods. It subsequently became an enterprise with the social responsibility to combine ecology with economic profit (Eyhorn, 2007). Presently, the bioRe India project is active in the Nimar Valley, involving 6,400 smallholders who cultivate 6,502 ha of cotton and produce more than 3,963 tonnes of seed cotton per year. An independent but related organization, the bioRe Association, runs a training centre offering education in organic farming and provides interest-free credit to farmers to develop infrastructure such as biogas facilities, vermicomposts, irrigation facilities and provisions for safe drinking water. bioRe Association also provides farmers with farm inputs such as seeds, deoiled castor, rock phosphate, botanical and microbial pesticides, etc., if needed, as well as vocational training and advisory services. bioRe enters into five-year purchase guarantee contracts with farmers, purchases the seed cotton at prevailing market rates and pays an up to 20 per cent price premium at the end of the season to farmers who have completed the three-year conversion period. In the first year, farmers receive the inputs on a credit basis, while in the following years input costs are adjusted with the price premium from the previous year. However, these supports are available only for the conversion and production of organic cotton. Farmers do not receive the same technical or market support to grow other crops under organic management. Panneerselvam et al. (2011a) surveyed 40 organic and 40 conventional small farmers from this case area. The results showed that organic cotton yields were on par with conventional cotton but organic cotton had lower costs of production and higher gross margins than did conventional cotton. Organic farms recorded lower yields for other crops such as wheat, resulting in lower overall farm productivity. However, this was compensated to some extent through greater use of intercropping. The farm gross margin was higher under the conventional system, the net margin turned out to be similar in the two systems due to a significant reduction of the input cost under the organic system. Therefore, small farmers improved food security by increasing the intercropping yield and reducing the cost of production.

The proportion of total farm production undertaken using organic practices often depends on the type and extent of institutional support available for resource-poor farmers. Private agribusiness companies, for example, often provide technical support to certify specific crops to attract the lucrative organic export market. Where resource-poor farmers lack knowledge and/or inputs to grow the full range of rotational crops using organic practices this may lead to mixed organic/conventional production. Moreover, a certified organic cash crop for export may divert factors of production away from food crop production and thereby reduce food self-sufficiency and food production. This was observed in Madhya Pradesh where organic farmers had technical and other support for growing cotton but not for other crops which led to lower overall farm production as well as lower wheat yield (Box 2.2). Similarly, in Uganda, reduced local food production was found among organic coffee farmers due to the expansion of coffee plantings at the expense of food crops such as maize and sweet potatoes (Bolwig and Odeke, 2007). However, organic farmers did improve their incomes due to increased yields and price premiums for cash crops. It is also important to note that resource-poor farmers do not necessarily lack the capacity to apply technologies and skills acquired through their participation in an organic export operation (e.g. bio-pesticides and organic soil fertility management) in food crop farming, thus raising food crop yields and/or reducing the cost of production.

Where NGOs provide the institutional infrastructure to support organic production a range of outcomes has been observed. Some NGOs operate similar models to private agribusiness focused on developing specialized value chains for certified organic produce. Others provide technical support for all crops grown on the farm and often the products are not certified and are sold to local markets (Panneerselvam et al., 2011b). Such a focus on farm agro-ecologies often leads to increased farm yield due to crop diversification and increased land-use intensity. Pretty et al. (2006) analysed 286 projects in 57 poor developing countries covering 37 million ha; they found that when agro-ecological practices covering a variety of systems and crops were adopted, average crop yields increased by 79 per cent. However, yield improvements did vary considerably. A quarter of all projects surveyed achieved yield increases of at least 100 per cent (relative yields >2.0) while half of all projects surveyed achieved yield increases of at least 18 per cent. A subset of the above study for Africa, consisting of 114 projects covering 2 million ha from 24 African countries, also demonstrated higher yields under organic farming – in many cases a doubling of yields – making an important contribution to increasing regional food security (UNEP-UNCTAD, 2008). The average crop yield increase was 116 per cent for all African projects, 128 per cent for projects in East Africa and 179 per cent for projects in Kenya. The major reasons for yield increase were reported as: (i) more efficient water use in both dryland and irrigated farming; (ii) improvements in organic matter accumulation in soils and carbon sequestration; and (iii) pest, weed and disease control emphasizing in-field biodiversity and reduced pesticide (insecticide, herbicide and fungicide) use.

Adoption of agro-ecological methods/organic farming techniques and efficient use of family labour in smallholding farms increase yield. Bolwig et al. (2009) observed each additional organic practice generates a 7 per cent increase of coffee yield on average in tropical Africa. Organic smallholders in Africa have doubled cashew nut harvest in responding to high demand (Forss and Lundström, 2005) through efficient use of family labour. Similarly, organic small farmers increased cauliflower yield by intensive use of family labour, whereas organic soybean farms were larger in size and not able to increase their yields to the same degree as small farmers of conventional soybean (Oelofse et al., 2010).

Food access

Food insecurity exists not only because of low production but also because people do not have adequate physical, social and economic access to food. Access to food can thus be improved either through increased household food production (discussed above) or by increasing incomes and purchasing power. The 2007 FAO conference stated that organic farming has the potential to improve the net incomes of farming households by reducing external input costs and/or enabling access to price premiums (Scialabba, 2007). The extent to which organic farming actually improves net income depends on reductions in

Table 2.2 Selected examples of comparisons between organic vs. conventional cash crop production in smallholder farms in Asia

	Rice, Philippines, 2000[1]		Soybeans, China[3]		Cotton, India[5]	
	Organic	Conventional	Organic	Conventional	Organic	Conventional
	US$ ha⁻¹		US$ ha⁻¹		Indian Rupees ha⁻¹	
Gross revenue	650[2]	564	1,088[4]	713	33,849[6]	26,078
Cash costs	39	118	}	}	7,796	9,334
Indirect costs[7]	149	155	305	640	2,369	2,650
Net revenue	462	290	78.3[4]	72.5	23,684	14,094
Yields, kg ha⁻¹	3,250	3,520	3,750	7,500	1,348	1,283
Labour use, man days ha⁻¹	49	52			190	181

Notes
1)   Mendoza, 2004.
2)   A 25 per cent price premium was obtained in certified organic.
3)   IFAD, 2005.
4)   Own calculations based on 2 years' prices given in IFAD (2005).
5)   Eyhorn et al., 2005. Numbers presented are averages of two years.
6)   Includes value of pulse intercrop and a 20 per cent price premium on organic.
7)   Mainly opportunity costs of own labour.

production costs, the availability of price premiums, access to markets and credit, and the relative yield of organic systems.

Both certified and non-certified organic products potentially receive price premiums depending on market access and conditions. Such premiums, along with reduced input costs, generally make organic farms more profitable than conventional farms even where yields are lower. However, the financial impact of removing synthetic fertilizer and pesticides from organic systems will depend on the extent to which they are substituted with commercially purchased organic manure and bio-pesticides or with labour. Table 2.2 thus summarizes several Asian studies comparing net income from organic and conventional production. These studies included, importantly, the opportunity cost of using family labour. With the exception of cotton in India, yields were lower under organic production. However, organic price premiums of 20 per cent or more saw higher gross returns in all studies. Reduced input costs were reported for all studies resulting in substantially higher net returns among Philippine rice and Indian cotton producers and marginally higher net returns among Chinese soybean producers. Not surprisingly, the largest increase in net revenue was found in cotton when organic systems both had higher yields and received a price premium. The modest increase in net income reported for Chinese soybean producers is explained by the dramatically lower yields under organic production for this particular crop. In a similar vein, Ramesh et al. (2010) surveyed various states of India and reported that organic farming, in spite of an average reduction in crop productivity of 9.2 per cent, improved farm net incomes by 22 per cent compared with conventional farming. This was mainly due to the availability of premium prices (20–40 per cent) for certified organic produce and a reduction in the cost of cultivation of 11.7 per cent. A small number of studies have reported higher production costs due to hired labour in organic farming although these have also found that in most cases the availability of price premiums compensated for extra costs and made organic farms more profitable. For example, organic oranges in Brazil achieved higher gross margins due to a 100 per cent increase in price, despite higher production costs and lower yields, whereas in China a 16 per cent price premium for organic soybean was enough to offset the low yield and high production cost (Oelofse et al., 2010).

We have calculated relative gross margin for cereals, oilseeds and other crops produced under organic and conventional systems from recently published papers from developing countries (Figures 2.4 and 2.5). The relative gross margin is the ratio of organic to non-organic gross margin reported by the studies. The results show that the gross margins of cereals, cotton, fruits and coffee were higher under organic production while the gross margins of oilseeds and vegetables were lower (Figure 2.4). However, the magnitude of increases in gross margin for the former crops was higher than the reductions in gross margin among the latter. For example, the gross margin for cereals and fruits increased by approximately 50 per cent under organic production and the gross margin for organic cotton and coffee by nearly 30 per cent. Several individual cereals and fruits achieved substantially higher relative gross margins; for example, a more than 100 per cent increase was reported for orange, maize, cashew nut and pepper, and between 50 and 100 per cent increase was reported for organic sorghum, mango, coffee, and sugarcane. Among individual crop categories that recorded lower relative gross margins only potato had a more than 30 per cent reduction under organic production while pigeon pea, peanut, soybean, cauliflower, vegetables, sugarcane, rice in Tamil Nadu, rice in Bangladesh and guava had 10–20 per cent reductions (Figure 2.5). All other crops had higher gross margin in organic systems. Further,

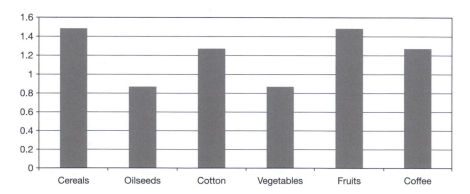

Figure 2.4 Relative gross margin for various commodities

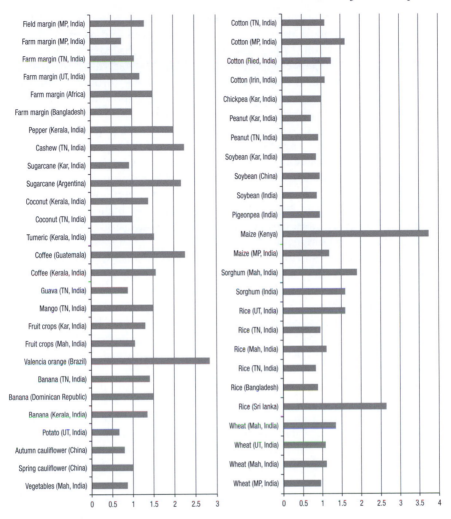

Figure 2.5 Relative gross margin of organic to conventional crops

the average total farm gross margin, or net return, for all crops grown on each farm in a single year was calculated to be 13 per cent higher under organic than under conventional production.

The majority of farms in developing countries that actively follow organic principles do so informally. They produce primarily for self-consumption and for local markets. They are not certified and generally they receive small or no price premiums. For example in India, Wai (2010) estimated that organic production amounts to about 1.62 million tonnes, out of which only 3 per cent was exported while the remaining 97 per cent of organic products were sold at domestic markets without or with small premiums. Informal or non-certified organic farms in Uttarakhand had 20 per cent higher gross margins due to the combination of lower input costs and higher yields (Panneerselvam et al., 2011a). Non-certified organic farms in Tamil Nadu were also found to profit due to lower costs of production (17 per cent), including farms that were in conversion to certified organic (Panneerselvam et al., 2011a). Organic rice farmers from Bohol, the Philippines, increased their incomes mainly through savings on fertilizers which translated into improved food security and reduced debt (Carpenter, 2003). However, the extent of cost reduction depends on the intensity of external input received before conversion. In India, the cost reduction was higher in converting high-input irrigated farms compared to low-input rainfed farms (Panneerselvam, 2011). Cost reductions were lower among the low-input farms of Uttarakhand compared with the high-input farms of Madhya Pradesh which had more opportunity to reduce expenditure on fertilizers and pesticides (Panneerselvam, 2011). However, in some cases, high-input farms had to purchase organic inputs, particularly during the conversion period, to meet the high nutrient requirement of crops.

Some agro-industrial firms extend their support of small farmers from production to the processing of organic products, creating opportunities to further increase farm income. Organic farmers in Uganda, for example, have increased their revenue not only by increasing yield and through access to price premiums but also by processing the coffee through a contract farming scheme (Bolwig et al., 2009). In total, an average increase in net coffee revenue of around 75 per cent was achieved, equivalent to 12.5 per cent of mean (total) household revenue. This was also due to incentives provided by the scheme to engage in coffee processing. Processing in the conventional market is costly in terms of time, labour and equipment and is an investment with uncertain returns. The existence of a price premium for scheme members may act to offset the risks associated with processing and, therefore, is likely to increase the extent to which farmers engage in these value-added activities. This organic scheme introduced clearer quality criteria and more transparent measurement of quality and volume than were available in the non-organic market, correcting this classic source of market failure of agricultural products in developing countries (Bolwig et al., 2009).

Above all, conversion to organic farming can improve farmers' access to food for a longer period and reduce the risk of indebtedness in case of crop failure. Ediriweera et al. (2007), for example, have recently shown that the number of farmers and the number of months those farmers face food insecurity in Sri Lanka is higher among conventional than among organic farmers. Bolwig and Odeke (2007) reported that African farmers significantly improved their food security due to the increase in coffee incomes which improved their ability to cover the household food deficit through the market. One indication of this was a reduction in (or elimination of) the length of the 'hunger period' (June–July) when households experience food scarcity, i.e. when there is no or little food in the house and no cash with which to buy it.

Food utilization

Food utilization means achieving an adequate diet, clean water, sanitation and health care to reach a state of nutritional well-being where all physiological needs are met. The food utilization dimension of food security has been linked with the organic farming principle of health (Vaarst, 2010). Organic agriculture has the potential to influence food utilization and human health by reducing the use of pesticides and improving nutritional efficiency by producing and supplying a greater diversity of food crops (Scialabba, 2007).

The major shift in cropping pattern towards green revolution crops such as rice and wheat has introduced a range of nutritional imbalances at the same time as it has increased calorific output (Singh, 2000). In India, for example, consumption of pulses declined from 72 grams per capita per day in 1956 to 42 grams per capita per day in 1990 and 33 grams in 2005 (Ministry of Finance, 2008), contributing to the persistence of protein calorie malnutrition among more than 50 per cent of the Indian population (INSA, 2009). The diversification of crops including pulses in the cropping pattern is a key step towards meeting dietary protein requirements. A number of studies show that organic farming promotes crop and other enterprise diversification including integration of more legumes which increases the availability of diversified food at the household level (Walaga and Hauser, 2005; Rasul and Thapa, 2004; Venugopalan et al. 2010; Scialabba and Hattam, 2002; Lyngbæk et al., 2001; Kilcher, 2006). Panneerselvam et al. (2011b) studied three states in India producing rice, mixed vegetables and cotton, where either 'avoidance of pesticides' or 'health benefits' was mentioned as a reason for conversion and experienced as advantages of organic farming after conversion. However, the focus of some market-oriented organic projects on single cash crops may compromise this diversity.

Another health benefit of organic agriculture is the absence of chemical pesticides with potential toxic effects on humans as well as ecosystems. Pesticide use is the cause of many health problems in developing countries due to a lack of knowledge and equipment for proper protection while spraying. The fact that farmers do not use such synthetic pesticides in organic systems is, thus, a potential health improvement. The actual improvement in this aspect in a given project or area depends, of course, on the pesticides used in the farming system before conversion. The farm families reduced their illness and associated treatment cost by avoiding use of pesticides in the farm in the Philippines which led to greater social benefits (Pretty et al., 2006). Similarly, in Karnataka, India, organic farmers reported not having experienced any feelings of illness after working in the organic rice fields, whereas more than half of the conventional farmers reported suffering at times from nausea and vomiting (IFAD, 2005). In Kerala, India, a number of farmers were hospitalized after local groundwater was contaminated with pesticide run-off from neighbouring tea estates (IFAD, 2005).

Food stability

Stability of the food supply is partly related to the agro-ecosystem resilience against climatic and biological perturbations and to environmental conditions that allow for sustainable food production and encourage productivity as well as to the economic conditions that allow for sustainable supplies at reasonable prices (Scialabba, 2007). In producing food, farmers can improve or degrade the environment and ecosystem services they depend on, thus impacting on the sustainability from a functional integrity perspective (Alrøe et al., 2006; Thompson, 2007). Hence, farming practices obviously have an important impact on the stability of food supplies over time. Practices that increase agro-ecosystem resilience by improving soil fertility and functional biodiversity generally contribute to food stability and so this dimension of food security is linked with the organic farming principles of health, ecology and care (Alrøe et al., 2006; Vaarst, 2010; see also Chapter 1). Organic principles encourage the use of crop mixtures and a generally high degree of so-called planned diversity (Power and Kenmore, 2002). As discussed by Perrings (2001), maintaining high biodiversity can be considered an insurance against catastrophically low yields due to climatic hazards or pests because it contributes to resilience. But, are these effects measurable in real farm situations?

Conversion to organic agriculture often has the effect of improved environmental sustainability through increased on-farm diversity of both cultivated and wild species, improved soil fertility and non-use of pesticides (Scialabba and Hattam, 2002; Halberg et al., 2006; Lockie and Carpenter, 2010). Documented environmental benefits of organic farming include lower resource use, improved nutrient recycling, reduced energy use, increased soil biological activity and preservation of biodiversity and landscape values (Refsgaard et al., 1998; Mäder et al., 2002; Bengtsson et al., 2005; Hole et al., 2005; Birkhofer et al., 2008; Lynch et al., 2011). These benefits are, however, mostly documented from the conversion of high-input farming systems dependent on agro-chemicals. In a context of non-industrialized agriculture, valid evidence of these benefits is more fragmented but does exist (Pretty, 2002; Scialabba and Hattam, 2002; Uphoff, 2002; Rasul and Thapa, 2004; Parrot et al., 2006). In Brazil, a few studies have documented environmental benefits of conversion to organic farming (Bellon et al., 2005). The environmental benefits of organic farming are discussed in Chapter 3.

A number of development projects have reported significant improvements in soil fertility and/or reduced erosion from introduction of agro-ecological methods such as agro-forestry and improved use of manure, mulching and other organic soil input (Araya and Edwards, 2006; Garrity et al., 2010; Christian Aid, 2011; Hivos and Oxfam Novib, 2009). As discussed in Chapter 4, not all types of organic agriculture increase the use of such methods and there is a need for more research and documentation on how conversion to certified and informal organic agriculture impacts on local food stability. This could go hand in hand with a further development of techniques and indicators to assess and document 'resilience' of farming systems, which is generally needed, not the least due to the necessity for climate change adaptation as discussed in Chapter 5.

Organic agriculture impact on other livelihood aspects

Organic farming projects have a positive effect on livelihood and capital assets but the form and degree depends on the types of organic farming as mentioned in the section 'Different forms of practising organic farming'. For example, certified organic farming in developing countries is often promoted by NGOs or private companies. They provide the associated farmers with training and technical advice on organic production and some extend interest-free credit (Box 2.2). Hence, farmers improve their skills, knowledge and capacity through participation and training, improve their health by avoiding fertilizers and agro-chemicals and have improved credit availability. Also, organic farmers improved their social networks through participating in training and group certification. The most important benefit of certified organic farming is improvement in the financial capital through price premium (detailed in the food access section). In Africa, improvement in financial capital among smallholders was demonstrated by improvement in housing, school attendance among farm children and investment in farming (EPOPA, 2008). Social responsibility has increased as a result of the group certification system (Wietheger, 2005).

Informal non-certified organic projects build on an agro-ecosystem approach and focus explicitly on improved management of natural resources for food security and environmental benefits. Such approaches tend to build *natural, human and social capital* (Bebbington, 1999) through training efforts and enhance the ability of people to address community and broader livelihood issues. This often results in diversified asset building if the intentions are fulfilled and may also involve significant amounts of technical and organizational training of both men and women. But it sometimes fails to improve *financial capital*. Informal non-certified organic farming in developing countries is often promoted by NGOs and development organizations. This type of organic farming often uses agro-ecological practices for nutrient management such as intercropping with pulses or application of farmyard manure, compost, vermicompost, neem cake or crop residues which leads to improvement in natural capital (of the soil). Also, non-certified organic farming seeks to increase the resilience (or reduce the susceptibility) of the farm family through diversification. The contribution of organic agriculture to natural capital is partly addressed in the section on food stability and more details are found in Chapters 3 and 4.

Organic farming often leads to improvements in human capital including increased knowledge and skills, reversed migration and more local employment. Organic agriculture's contribution to human health is discussed in the section on food utilization and some detailed aspects of human and social capital and power relations are discussed in Chapter 6. Some of the documented benefits of organic production on human capital include the return of families to rural areas to take up labour opportunities created by organic projects, for example in the Guinope and Cantarranas regions of Honduras (Bunch, 2000). According to a survey carried out in China (Greenpeace, 2003), the employment rate (work load) in organic farming can be 25–40 per cent higher compared with conventional production and may especially involve more labour from women. This corresponds with assessments by IFAD (2005) which found 30–40 per cent higher labour use in organic rice and tea crops in China and India. The increased labour demand might pose a risk that smallholder farmers choose to involve their children more in farm work at the cost of schooling. However, no such examples have been reported, to our knowledge, in the literature. On the other hand, many projects promoting organic agriculture include training for farmers which may increase willingness to send their children to school. The increased income from certified organic cash crops could also increase the possibilities for poor households to keep children in school. In a survey of resource-poor households in the Philippines, Mendoza (2004) reported that the significant reduction of cash capital expenses from approximately 118 US$ per ha in conventional rice to US$39 in organic rice contributed to the organic families' ability to pay school fees (coincidentally due also at the start of the cropping season).

Conclusions

Fully understanding the contribution of organic agriculture to livelihoods and food security requires studies that deal in greater detail with the relationships between livelihood strategies, total farm production, household income, household food intake, hunger status, indebtedness, nutritional status, education attainment, and so on. While few studies are able to capture such a level of detail, there is convincing evidence that adoption of organic agriculture can have a positive effect on smallholders' food security and livelihoods in developing countries. The extent to which organic agriculture improves food security and household economies depends on the interaction of several variables including, most notably, the characteristics of organic agriculture projects (e.g. certified or non-certified organic, export or local markets, etc.), the types of crops and growing conditions (e.g. irrigated or rainfed, specialized or diversified, intensive or extensive, etc.) and the relative cost and availability of synthetic and non-synthetic inputs. Organic agriculture has shown considerable potential to improve smallholders' food security by increasing food production in low-input areas under both formal and informal organic systems.

Certified organic farming has a number of positive effects on livelihood assets, primarily in the improvement of financial capital (through price premiums and reduced production costs), improved human capital including increased knowledge and skills, social capital through group certification and schooling of children through increased income. However, concern remains over the impacts of certified organic schemes that focus only on specific crops grown either on part of a landholding or at the expense of more diversified rotations. Moreover, certifying companies provide technical support for specific crops to attract the lucrative organic export market, but farmers are lacking knowledge and inputs to grow the full range of rotational crops that are imperative in a well-functioning organic production system. Thus, one of the biggest areas of potential lies in establishing organic markets for rotational crops in developing countries. If the promoting institutions are focused only on cash crops for export then there is a chance of diverting land from the production of food crops to cash crops which will have a negative impact on food availability and food prices at both micro and macro levels. Therefore, the promoting institutions in developing countries should promote diversified farming systems, integrating livestock and food crops with cash crops to increase food availability and income, and reduce the risk in case of crop failure. For this, local organic markets should be organized to meet the supply of any marketable surplus of food crops grown in the cash crops area.

Informal organic farming has been shown to increase natural, human and social capital through training and capacity building, improve health due to production of diverse food and avoidance of pesticides, and increase knowledge and skills. Informal non-certified organic farming is also increasing the income of smallholders by reducing the cash costs of production. This is enhanced where farmers have access to local markets with small premiums for their produce, again assisting farmers to access other livelihoods and send their children to school. While the productivity of such systems compares more than favourably with conventional farming systems in developing countries it is highly likely that poor management and inadequate access to inputs depresses the average performance of conventional farms. Matching or moderately exceeding this level of performance is not therefore adequate. More research and institutional support are required to increase organic yields and market access to continue improving food security and livelihoods. Moreover, informal non-certified organic farming has often been promoted and supported by NGOs and development organizations and this support was available only for certain periods due to constraints of project funding and thus it is important that a long-term strategy for the support is developed.

References

Alrøe, H.F., Byrne, J. and Glover, L. (2006) 'Organic agriculture and ecological justice: ethics and practice', in N. Halberg, H.F. Alrøe, M.T. Knudsen and E.S. Kristensen (eds) *Global Development of Organic Agriculture: Challenges and Prospects,* CABI International, Wallingford, pp. 75–112

Araya, H. and Edwards, S. (2006) 'The Tigray Experience. A success story in sustainable agriculture', *Third World Network,* Environment & Development Series, 4, p. 44

Badgley, C., Moghtader, J., Quintero, E., Zakem, E., Chappell, M.J., Avilés-Vàzques, K., et al. (2007) 'Organic agriculture and the global food supply', *Renewable Agriculture and Food Systems,* 22, 2, pp. 86–108

Bebbington, A. (1999) 'Capitals and capabilities: a framework for analyzing peasant viability, rural livelihoods and poverty', *World Development,* 27, 12, pp. 2021–2044

Bellon, S., Santiago de Abreu, L. and Valarini, P.J. (2005) 'Relationships between social forms of organic horticultural production and indicators of environmental quality: a multidimensional approach in Brazil', in U. Köpke, et al. (eds) *Researching Sustainable Systems*, Proceedings of ISOFAR, Adelaide, Sept.

Bengtsson, J., Ahnström, J. and Weibull, A.C. (2005) 'The effects of organic agriculture on biodiversity and abundance: A meta-analysis', *Applied Ecology*, 42, pp. 261–269

Birkhofer, K., Bezemer, T.M., Bloem, J., Bonkowski, M., Christensen, S., Dubois, D., et al. (2008) 'Long-term organic farming fosters below and aboveground biota: implications for soil quality, biological control and productivity', *Soil Biology and Biochemistry*, 40, pp. 2297–2308

Bolwig, S. and Odeke, M. (2007) 'Household food security effects of certified organic export production in tropical Africa – a gendered analysis', EPOPA publication, Bennekom, The Netherlands

Bolwig, S., Gibbon, P. and Jones, S. (2009) 'The economics of smallholder organic contract farming in tropical Africa', *World Development*, 37, pp. 1094–1104

Bunch, R. (2000) 'More productivity with fewer external inputs', *Environment, Development and Sustainability*, 1, 3–4, pp. 219–233

Carpenter, D. (2003) 'An investigation into the transition from technological to ecological rice farming among resource poor farmers from the Philippine island of Bohol', *Agriculture and Human Values*, 20, pp. 165–176

Chambers, R. and Conway, G. (1992) 'Substantial rural livelihoods: practical concepts for the 21st century', Discussion Paper 296. Institute for Development Studies, University of Sussex, Brighton

Christian Aid Report (2011) 'Healthy harvests: the benefits of sustainable agriculture in Africa and Asia', available at: www.christianaid.org.uk/images/Healthy-Harvests-Report.pdf

DES (2009) 'Agricultural statistics at a glance 2009', Directorate of Economics and Statistics, Ministry of Agriculture, Government of India, available at http://eands. dacnet. nic.in/At_Glance_2009.htm (Accessed on April 26, 2010)

DFID (1999a) 'Sustainable Livelihoods and Poverty Elimination', Department for International Development, London

DFID (1999b) 'Sustainable Livelihoods Guidance Sheets', Department for International Development, London, available at www.livelihoods.org/info/info_guidancesheets. html (accessed December 4, 2011)

Dorward, A. and Poulton, C. (2008) 'The global fertilizer crisis and Africa', Briefing Paper, Future Agricultures Consortium. Available at www.future-agriculture.org

Ediriweera, E.S., Lieblein, G., Bandara, J.M.R.S. and Francis, C. (2007) 'Organic and conventional farming systems contribution to household food security in Sri Lanka', in *Papers Submitted to the International Conference on Organic Agriculture and Food Security*, FAO, Rome, Italy, 3–5 May, pp. 119–120

EPOPA (2008) 'Organic exports – a way to a better life?' Available at www.grolink.se/epopa/publications/Epopa-end-book.pdf

Evans, A. (2009) 'The feeding of the nine billion: global food security for the 21st century', Available at www.chathamhouse.org.uk/publications/papers/view/-/id/694/ (accessed on January 24, 2009)

Eyhorn, F. (2007) 'Organic farming for sustainable livelihoods in developing countries: the case of cotton in India', PhD dissertation, Department of Philosophy and Science, University of Berne, Switzerland, available at http://www.sb.unite.ch/download/eldiss/Obeyhorn_f.pdf

Eyhorn, F., Ramakrishnan, M. and Mäder, P. (2007) 'The viability of cotton-based organic farming systems in India', *International Journal of Agricultural Sustainability*, 5, pp. 25–38

FAO (Food and Agricultural Organization) (2005) 'FAOSTAT database', http://faostat. fao.org (accessed on January 15, 2009)

FAO (Food and Agricultural Organization) (2008a) 'The state of food insecurity in the world'. Available at www.fao.org/docrep/011/i0291e/i0291e00.htm (accessed on January 15, 2009)

FAO (Food and Agricultural Organization) (2008b) 'An introduction to the basic concepts of food security', FAO Food Security Programme. Available at www.food sec.org/docs/concepts_guide.pdf

FAO (Food and Agricultural Organization) (2010) 'The state of food insecurity in the world'. Available at www.fao.org/docrep/013/i1683e/i1683e.pdf (accessed on July 10, 2010)

Forss, K. and Lundström, M. (2005) 'An evaluation of the program "Export Promotion of Organic Products from Africa"', phase II, Sida, available at www. epopa.info

Garrity, D.P., Akinnifesi, F.K., Ajayi, O.C., Weldesemayat, S.G., Mowo, J.G., Kalinganire, A. et al. (2010) 'Evergreen agriculture: a robust approach to sustainable food security in Africa', *Food Security*, 2, pp. 197–214

Greenpeace (2003) 'The comparison research on the China organic agriculture development and its benefit', Project Report

Halberg, N. and Kristensen, I.S. (1997) 'Expected crop yield loss when converting to organic dairy farming in Denmark', *Biological Agriculture and Horticulture*, 14, pp. 25–41

Halberg, N., Peramaiyan, P. and Walaga, C. (2009) 'Is organic farming an unjustified luxury in a world with too many hungry people?', in H. Willer, and L. Klicher (eds) *The World of Organic Agriculture, Statistics and Emerging Trends 2009*, FiBL and IFOAM, Frick and Bonn, pp. 95–101

Halberg, N., Rosegrant, P., Sulser, T., Knudsen, M.T. and Høgh-Jensen, H. (2006) 'The impact of organic farming on food security in a regional and global perspective', in N. Halberg, M.T. Knudsen, H.F. Alrøe and E.S. Kristensen (eds), *Global Development of Organic Agriculture: Challenges and Prospects*, CABI Publishing, Wallingford, pp. 277–322. Available at http://ecowiki.org/GlobalPerspective/ ReportOutline

Hassanali, A., Herren, H., Khan, Z.R., Pickett, J.A. and Woodcock, C.M. (2008) 'Integrated pest management: the push–pull approach for controlling insect pests and weeds of cereals, and its potential for other agricultural systems including animal husbandry', *Philosophical Transactions of the Royal Society B*, 363, pp. 611–621

Hivos and Oxfam Novib (2009) 'Biodiversity, livelihoods and poverty: lessons learned from 8 years of development aid through the Biodiversity Fund', The Hague, Netherlands

Hole, D.G., Perkins, A.J., Wilson, J.D., Alexander, I.H., Grice, P.V. and Evans, A.D. (2005) 'Does organic farming benefit biodiversity?', *Biological Conservation*, 122, pp. 113–130

IFAD (2002) 'Thematic evaluation of organic agriculture in Latin America and the Caribbean', Evaluation Committee Document, 290404, International Fund for Agricultural Development, Rome

IFAD (2005) 'Organic agriculture for poverty reduction in Asia: China and India focus', *Thematic Evaluation*, Report 1664, International Fund for Agricultural Development, Rome, p. 80

IFOAM (2004) *Position on the full diversity of Organic Agriculture*, available online: http://www.ifoam.org/press/positions/full-diversity_organic-agriculture.html

INSA (2009) 'Nutrition security for India: issues and way forward', Indian National

Science Academy, Symposium report, available at http://typo3.fao.org/fileadmin/user_upload/fsn/docs/Symposium_Report_NutritionSecurityIndia.pdf (accessed on August 8, 2010)

Key, N. and Runsten, D. (1999) 'Contract farming, smallholders and development in Latin America', *World Development*, 27, pp. 381–401

Kilcher, L. (2006) 'The contribution of organic farming to sustainable development', in H. Willer and M. Yussefi (eds) *The World of Organic Agriculture 2006 – Statistics and Emerging Trends,* IFOAM, Bonn, pp. 91–95

Lockie, S. and Carpenter, D. (2010) 'Agrobiodiversity and sustainable farm livelihoods: policy implications and imperatives', in S. Lockie and D. Carpenter (eds) *Agriculture, Biodiversity and Markets: Livelihoods and Agroecology in Comparative Perspective*, Earthscan, London

Lynch, D.H., Halberg, N. and Bhatta, G. (2011) 'Environmental impact of organic agriculture: a review with special reference to North America', *CAB Reviews: Perspectives in Agriculture, Veterinary Science, Nutrition and Natural Resources*, 7, 10, pp. 1–17

Lyngbæk, A.E., Muschler, R.G. and Sinclair, F.L. (2001) 'Productivity and profitability of multistrata organic versus conventional coffee farms in Costa Rica', *Agroforesty Systems,* 53, pp. 205–213

Mäder, P., Fleisbach, A., Dubois, D., Gunst, L., Fried, P. and Niggli, U. (2002) 'Soil fertility and biodiversity in organic farming, *Science*, 296, pp. 1694–1697

Mendoza, T.C. (2004) 'Evaluating the benefits of organic farming in rice agroecosystems in the Philippines', *Journal of Sustainable Agriculture,* 24, 2, pp. 93–115

Ministry of Finance (2008) 'Economic survey, 2007–08', Ministry of Finance, Government of India, available at http://indiabudget.nic.in/es2007-08/chapt2008/chap72.pdf

Oelofse, M, Høgh-Jensen, H., Abreu, L.S., Almeida, G.F., Hui, Q.Y., Sultan, T. and de Neergaard, A. (2010) 'Certified organic agriculture in China and Brazil: market accessibility and outcomes following adoption', *Ecological Economics*, 69, 9, pp. 1785–1793

Panneerselvam, P. (2011) 'Improving smallholder's food security through organic agriculture in India', PhD thesis, Aarhus University, Denmark

Panneerselvam, P., Hermansen, J.E. and Halberg, N. (2011a) 'Food security of small farmers: comparing organic and conventional systems in India', *Journal of Sustainable Agriculture*, 35, pp. 48–68

Panneerselvam, P., Halberg, N., Vaarst, M. and Hermansen, J.E. (2011b) 'Indian farmers' experience with and perception of organic farming', *Renewable Agriculture and Food Systems* 27(2), pp. 157–169

Parrot, N., Olesen, J. E. and Høgh-Jensen, H. (2006) 'Certified and non-certified organic farming in the Developing World', in N. Halberg, M.T. Knudsen, H.F. Alrøe and E.S. Kristensen (eds) *Global Development of Organic Agriculture: Challenges and Prospects,* CABI Publishing, Wallingford, pp. 153–176, available at http://ecowiki.org/GlobalPerspective/ReportOutline

Perrings, C. (2001) 'The economics of biodiversity loss and agricultural development in low-income countries', in D.R. Lee and C.B. Barrett (eds) *Tradeoffs or Synergies? Agricultural Intensification, Economic Development and the Environment*, CABI International, Wallingford

Power, A.G. and Kenmore, P. (2002) 'Exploiting interactions between planned and unplanned diversity in agroecosystems: what do we need to know?' in H. Uphoff (ed.) *Agroecological Innovations*, Earthscan, London, pp. 223–242

Pretty, J. (2002) *Agri-Culture: Reconnecting People, Land and Nature*, Earthscan, London

Pretty, J.N., Morison, J.I.L. and Hine, R.E. (2003) 'Reducing food poverty by increasing agricultural sustainability in developing countries', *Agriculture, Ecosystems and Environment*, 95, pp. 217–234

Pretty, J., Noble, A., Bossio, D., Dixon, J., Hine, R.E., Penning de Vries, P. and Morison, J.I.L. (2006) 'Resource conserving agriculture increases yields in developing countries', *Environmental Science and Technology*, 40, 4, pp. 1114–1119

Ramesh, P., Panwar, N.R., Singh, A.B., Ramana, S., Yadav, S.K., Shrivastava, R. and Rao, A.S. (2010) 'Status of organic farming in India', *Current Science*, 98, pp. 1190–1194

Rasul, G. and Thapa, G.B. (2004) 'Sustainability of ecological and conventional agricultural systems in Bangladesh: an assessment based on environmental, economic and social perspectives', *Agricultural Systems*, 79, pp. 327–351

Refsgaard, K., Halberg, N. and Kristensen, E.S. (1998) 'Energy utilization in crop and livestock production in organic and conventional livestock production systems', *Agricultural Systems*, 57, pp. 599–630

Schmidhuber, J. and Tubiello, F.N. (2007) 'Global food security under climate change', *Proceedings of the National Academy of Science of the United States of America*, 104, pp. 19703–19708

Scialabba, N.E. (2007) 'Organic agriculture and food security', paper presented at International Conference on Organic Agriculture and Food Security', 3–5 May, Food and Agriculture Organization of the United Nations, Italy (OFS/2007/5), available at www.fao.org/organicag/ofs/index_en.htm

Scialabba, N.E. and Hattam, C. (eds) 2002 'Organic agriculture, environment and food security', Environment and Natural Resources Service Sustainable Development Department, FAO, Rome

Selvaraj, N. and Mohandas, B. (2008) 'Development of integrated organic farming system and marketing of hill horticultural crops – a case study in Nilgris', in M.K. Menon, Y.S. Paul and N. Muralidhara (eds) *Global Organic Agribusiness: India Arrives*, Westville Publishing House, New Delhi, pp. 36–42

Singh, R.B. (2000) 'Environmental consequences of agricultural development: case study from the Green Revolution state of Haryana, India', *Agriculture, Ecosystem and Environment*, 82, pp. 97–103

Thompson, P.B. (2007) 'Agricultural sustainability: what it is and what it is not', *International Journal of Agricultural Sustainability*, 5, 1, pp. 5–16

UNEP-UNCTAD (2008) 'Organic Agriculture and Food Security in Africa' (UNCTAD/DITC/TED/2007/15), United Nations Conference on Trade and Development; United Nations Environment Programme, available at http://www.unep-unctad.org/cbtf (accessed on February 25, 2009)

Uphoff, N. (2002) *Agroecological Innovations – Increasing Food Production with Participatory Development*, Earthscan, London

Vaarst, M. (2010) 'Organic farming as a development strategy: who are interested and who are not?' *Journal Sustainable Development*, 3, pp. 38–50

Venugopalan, M.V., Rajendran, T.P., Chandran, P., Goswami, S.N., Challa, O. and Damre, P.R. (2010) 'Comparative evolution of organic and non-organic cotton production systems', *Indian Journal of Agricultural Sciences*, 80, pp. 287–292

Wai, O.K. (2010) 'Organic Asia 2010', in H. Willer and L. Klicher (eds) *The World of Organic Agriculture, Statistics and Emerging Trends 2010*, FiBL and IFOAM, Frick and Bonn, pp. 122–127

Walaga, C. and Hauser, M. (2005) 'Achieving household food security through organic agriculture, lessons from Uganda', *Journal für Entwicklungspolitik,* 21, 3, pp. 65–84

Warning, M. and Key, N. (2002) 'The social performance and distributional consequences of contract farming: an equilibrium analysis of the Arachide de Bouche programme in Senegal', *World Development,* 30, pp. 255–263

Wietheger, L. (2005) 'Impacts of the development programme EPOPA on peasants' livelihoods in Kagera/Tanzania', Wageningen University, Msc thesis

World Food Summit (1996) 'Rome Declaration on World Food Security and World Food Summit, Plan of Action', FAO, Rome

Case study 1
Food security obtained through Farmer Family Learning Group approaches

Description of a project between Organic Denmark, NOGAMU and SATNET Uganda

Inge Lis Dissing, Jane Nalunga,
Thaddeo Tibasiima, Aage Dissing and Mette Vaarst

Introduction

In 2008 Organic Denmark and NOGAMU (National Organic Agricultural Movement of Uganda) initiated a joint project aiming at improved livelihood and food security. SATNET is a NOGAMU member and an umbrella organization consisting of about 50 local member organizations in western Uganda, and it was selected for the pilot project. SATNET had supported the development and adaption of agro-ecological farming using different approaches to training for more than ten years.

The aim of this project in the Rwenzori Region was to develop an approach for farmer families and local communities to improve their livelihood and environments based on common learning and efforts in farmer group inter-actions. We took the starting point in the concepts of Farmer Field Schools (FFS), which is a concept that can be practised in very many ways. We have used a very flexible approach with great success, and we started calling the groups 'Farmer Family Learning Groups' (FFLG) for several reasons. On many occasions, we found that we were not following the FFS principles accurately. We involved the farms of all participating group members as learning sites, did not work with demonstration farms or demonstration plots, and we did not focus on one enterprise. In other words, we did not follow the curricula of any FFS manual, and training only took place on request from the group. Second, we increasingly saw the importance of involving whole families in the group, in the activities and the discussions and planning, and this became a focus area. Third, we emphasized the active participation and involvement of everybody – which gradually replaced the 'school' with the concept of 'learning'. An FFLG has a continuous life and goes through phases of establishment to becoming a mature group, which constantly develops by taking in new activities, based on the wishes and ambitions of the group members.

The project team understands a Farmer Family Learning Group as a group of households or farmer families which – with the help of a facilitator – get together to support and help each other. The idea is to solve their common problems through working, developing and learning together and, when relevant, marketing their produce. An FFLG is an approach that should be owned by the group members, and each group will be shaped and formed by their needs, which change over time. We believe that the group members' ownership over the group and its approaches and activities is the sole pathway to sustainable groups. We attempt to enable facilitators to form and facilitate groups based on an identification of what the members want and need.

The beauty of the approach – like many other empowering group approaches – is its strong and clear foundation on values including respect, trust, equality, common learning, building up human and social capital and knowledge which is relevant and meaningful to each participant and learner, as well as probably the most important value: ownership.

Food security and income generation

Many families were food insecure in part of the year, before the project started. According to a baseline study including four focus group interviews, the type of food in a large part of the area was described as quite unbalanced, with much carbohydrate-rich food and little protein and vegetables.

The main focus of this project was improved food security in the communities of West Uganda. We found that organic agriculture, understood as farming according to the principles of health, care, ecology and fairness, and the conscious use of agro-ecological methods, is a highly relevant approach for improved food security in this region. The conscious use of agro-ecological methods and practices requires knowledge and skills. Some traditional farming methods are also relevant for organic farming, and this requires knowledge about these methods as well. Whenever knowledge is required, it is relevant to create situations where knowledge can be exchanged, developed and debated. The pool of knowledge and experience in a group of farmers is tremendous, and when exhausted, they can agree to find ways to bring in new inspiration.

The practices of agro-ecological farming require labour, e.g. for trench digging, mulching, weeding and compost production. In all FFLGs, people get together and help each other with these big tasks. For example, by working together on each other's plots, 20 to 30 people can make a family's banana plantation well maintained in a few hours. This is much more encouraging than working the same amount of time alone on one's own land. The social aspect of working together in a group adds to building up social capital in a local community.

Based on a common analysis of the challenges and conditions, as explained above, the project team formulated the following requirements to direct the focus and efforts for an initiative to form farmer groups:

- Whole families must be involved. The work and responsibilities on a farm are distributed between family members. The children need to be involved, because they are the future of farming, and for this they need to learn and experience farming; there seems to be a worldwide gradual disconnection between the lives of children and youth on one side, and agricultural practice and farming on the other side.
- Building social capital is a main focus of the group formation. Appropriate time must be given for each group to identify their focus and build networks. They should be sufficiently guided in this by the external facilitator.
- The facilitators should be educated to allow the process of letting the group find their own pathway. The way in which each group will work should be identified in a dialogue between the facilitator and group members.
- The local organizations found that sustainability of the groups is more likely assured when they work with their own facilitator chosen amongst them, rather than a facilitator who worked from an organization, even when locally based. We therefore chose to educate so-called 'external facilitators' found from within the local organizations. After this, the external facilitators could initiate groups, let the group select a so-called 'internal facilitator', who would then be guided by the external facilitator. The internal facilitator would be a part of the local community and would therefore stay with the group.
- All participating farmer families should be committed to open up and expose their own farm to the group, and to respectfully help all their fellow farmer families in the group.

In this first phase 35 Farmer Family Learning Groups were established and they are still active, each composed of 8 to 25 families or roughly 4,400 individuals.

Case stories from the project

The farmers in one of the groups near Kasese had very small gardens and farm land in general (a quarter to two and a half acres). They rapidly improved their gardens especially in terms of diversification. They included several species of vegetables and this helped them to be able to make a more balanced diet for their families. In addition, they improved their compost making and focused on utilizing a water trench which was running through their village from a stream; both initiatives supported the increased and diversified vegetable production.

> Now we are food secure and as women we have started earning income from marketing our surplus vegetables. There is more collaboration and we have started our own Saving and Credit Scheme.

In another group close to Bundibugyo, food security was established in another way. One woman explained:

> There is a big improvement in my garden since Farmer Field School. There is cocoa – maybe when you go you can see also cassava, but . . . you cannot go anywhere without seeing cocoa. Before, we only sold cocoa and got the money. I have now increased the income or my daily earnings because we don't buy food. I have now improved the house and we now grow food everywhere.

With this progress she is able to secure food without depending on the cocoa prices. We can see it when we go there – she has completed her house, and there are animals and food crops, besides the organic certified cocoa.

All family members including the children generally joined the work in the groups. At one of the monitoring meetings in a group south of Fort Portal, it was actually an 11-year-old child who showed us around her family farm. She was obviously very well informed and knowledgeable on all the farming practices. This FFLG was established quite early in the project and was launched as a so-called demonstration project, recognized by the public authorities at the district level. It was also promoted as a learning place for other farmer groups in the area to visit to learn about possible farm improvements.

In two of the groups, the external facilitator left the group very early after the initiation of the FFLG, and an internal facilitator was appointed but had had very little training. In one case, this happened in time to allow him to participate in a four-day post-training course for the external facilitators. This exposed him to very many discussions about facilitation and the idea of the farmer groups. The group consisted of 17 farmer families (32 group members) in a mountainous area northwest of Fort Portal. They developed their farms focusing on control of soil erosion, mainly by digging trenches, mulching and planting trees. This was followed by the selection of highly diversified cash crops like Irish potatoes, garlic, beans, peas (a special type that matured in 60 days), coffee, onions, passion fruit and barley for beer production. They also had goats and chickens. The group has bought land to construct a store room for cash crops for joint marketing.

Evaluation

At the end of the project period all external and internal facilitators had a final evaluation. They had discussed with their FFLG the most significant change during the two-year period, and through a group interactive process they selected the following headline as the most significant change:

> *Interrelationships through working together.*

They also concluded that *improved food security, focused planning of the land use, higher productivity, social responsibility and group micro-finance were results of the stronger relationships in families as well as in local communities.*

Figure 1 Local farmer demonstrating her improved banana plantation

Figure 2 A family learning group

Figure 3 Farmers discussing plans for improved cropping systems based on their own
farm maps

3 Globalization of organic food chains and the environmental impacts

John E. Hermansen,
Marie Trydeman Knudsen and
Christian Schader

Introduction

During the last ten years the organic market has increased several fold and now exceeds 50 billion US$ (Willer and Kilcher, 2010). At the same time organic farming and food systems have shown a rapid development towards increasing global trade with organic products. The characteristic development has been the rise in demand and consumption in Europe and North America, while the rise in organic production and in number of organic farmers was in the South – the South in particular supplying tropical products and out-of-season products for Northern consumers. A driving force for this development – apart from the demand in the North – has been the establishment of frames for global trade with organic products.

The growth in the major organic markets in Europe and North America offers export opportunities to developing countries, and can be seen as a development pathway with economic and environmental benefits for the farmers, regions and countries in the South (e.g. Twarog, 2006; FAO, 2001). However, the growing global trade with organic products, which also implies long-distance trade, has given rise to a debate on the sustainability of this development, especially with regard to global warming and the carbon footprint of organic products (e.g. Soil Association, 2007; Rigby and Bown, 2003). In this debate, the long-distance transport resulting from global trade with organic products is said to be challenging one of the four basic principles of organic agriculture (see Chapter 1). The principle of ecology states that 'Those who produce, process, trade or consume organic products should protect and benefit the common environment including landscapes, climate, habitats, biodiversity, air and water.' It is further stated that 'Inputs should be reduced by reuse, recycling and efficient management of materials and energy' (IFOAM, 2005). Long-distance transport is seen as challenging this principle of efficient use of energy and adds to global climate change. Furthermore, the principle of recycling nutrients is challenged by the global as well as regional trade, with organic products supplying metropolitan centres in the North and South (Refsgaard et al., 2006).

In addition to the debate on environmental impacts, it is also argued that the prerequisites for the globalized market for organic products limits small-scale farmers in entering the market due to the costs of certification (Barrett et al., 2002) and demands for large and stable quantities to be delivered (Kledal, 2009; Blanc, 2009), which tend to favour large-scale organic farms. This could be seen as challenging the principle of fairness (IFOAM, 2005; see also Chapter 6).

This leads to the question of what kind of agricultural development urban consumers in the North and South support when choosing organic products and what are the environmental implications. How much does the long-distance transport associated with importing organic products contribute to greenhouse gas emissions – and how do the organic systems in the producing countries affect the environment compared to conventional systems?

This calls for transparency in the organic food chains and is relevant in order to secure long-term market relationships for smallholder farmers. Thus, this chapter focuses on environmental issues, while other important aspects are dealt with in other chapters, e.g. aspects of food security are dealt with in Chapter 2 and certain aspects of power inequalities are addressed in Chapter 6. Following an overview of the global market in the following section, this chapter examines the existing evidence on the environmental impacts related to organic and conventionally produced food and then considers the contribution from transport in a number of relevant situations. Focus is on greenhouse gas emissions which is an important impact category in transport as well as production of food, and is also correlated with other important environmental impact categories. The chapter briefly addresses impact categories that pose more difficulties to life cycle assessment (LCA) such as biodiversity and draws conclusions in the final section.

Major routes in food import from developing countries to Europe and North America

The global markets for organic food and drink have grown remarkably during the last decade, and global sales of organic products have more than doubled from 2003 to 2008 (Figure 3.1).

The major markets for organic food and drink are still Europe and North America, which account for 97 per cent of global revenues (Figure 3.1). The Scandinavian and Alpine countries are the largest consumers of organic foods, with an organic market share of more than 4 per cent of total food and drinks sales in Denmark, Austria, Switzerland and Sweden (Willer and Kilcher, 2010). The Danish consumers are the world's leading buyers of organic food (Willer and Kilcher, 2010) – and the increased market to a high degree is covered by the import of organic foods to Denmark. In addition to Europe and North America, the Asian markets for organic food and drink are also growing, especially in the more affluent countries, such as Japan, South Korea, Taiwan, Singapore and Hong Kong. Furthermore, domestic markets are developing in some major cities in Latin America (Willer and Kilcher, 2010).

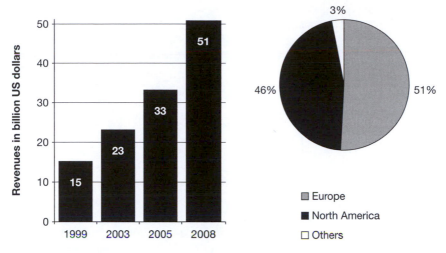

Figure 3.1 Global market growth for organic products (left) and distribution of global
revenues by region in 2008 (right)

Source: *Organic Monitor*, 2010.

These markets offer good prospects for suppliers of organic products and vast
areas of agricultural land are located far away from the major markets.
Consequently, one-third of the world's organically managed land is located in
developing and the BRICS countries, with Argentina, China and Brazil having
the most organic land (Willer and Kilcher, 2010). In the following discussion,
some characteristics of the organic agricultural product flow from Latin America,
Asia and Africa to Europe and North America are described.

Figure 3.2 illustrates the import of some organic agricultural products to
Denmark in the period 2006–2010. Interestingly, the rapidly growing imports
to Denmark consist mainly of fruit and vegetables, sugar, rice and cereals,
whereas the export from Denmark is mainly animal-based products. The same
pattern is seen in the UK, where 82 per cent of the organic fruit and vegetable
sales were met by imports in 1999, while only 5 per cent of the organic meat
sales were imported (Barrett et al., 2002). This can be partly explained by the
short season for fruit and vegetables in the North. Apart from this extra need
for fruit and vegetables in the North, the organic import pattern of Denmark
from tropical countries might also apply to the rest of Europe and North
America. The main share of the organic imports to Denmark comes from
Europe, followed by South America, Asia and Africa (StatBank Denmark, 2010).

Latin America holds 23 per cent of the world's organic agricultural land with
most of the production destined for the export markets (Willer and Kilcher,
2010). The main export products from Latin America are coffee (Mexico, Peru,
etc.), cocoa (Dominican Republic, Ecuador, etc.), sugar (Brazil, Paraguay, etc.)
and fruits such as bananas (Dominican Republic, Ecuador, etc.), apples and pears
(Argentina, etc.) and citrus products (Brazil, Mexico, etc.) in addition to other

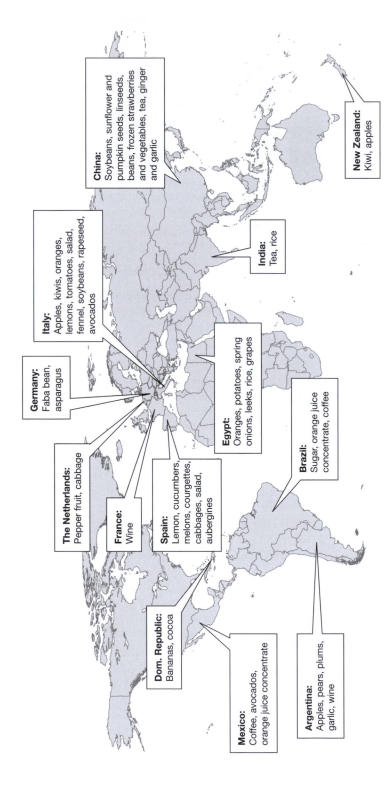

Figure 3.2 Import of selected organic food, drinks and fodder to Denmark in the period 2006–2010

Source: Based on information from supermarkets, retailers and Danish official statistics (StatBank Denmark, 2010).

products such as wine (Chile, Argentina, etc.) and avocado (Mexico, etc.) (Garibay and Ugas, 2010; Barrett et al., 2002). Latin America provides three-quarters of the world's organic coffee consumed mainly in the USA and Europe (Giovannucci and Pierrot, 2010). Argentina holds the largest organic agricultural land in Latin America followed by Brazil, Uruguay and Mexico, but vast areas especially in Argentina are permanent grasslands. In contrast, the Dominican Republic is one of the countries in Latin America with the highest percentage of total agricultural land under organic management (6.3 per cent) (Willer and Kilcher, 2010).

Asia has 9 per cent of the world's organic agricultural land, which is mainly located in China (1.9 mill ha) and India (1.1 mill ha) (Willer and Kilcher, 2010). China has become a global source of organic products with the primary driver being trade and export (Yin et al., 2010). The main export products from China are grains, beans, fruit and vegetables, accounting for 90 per cent of the organic exports (Sheng et al., 2009). From India, the main organic export products are cotton, tea and rice in addition to dried fruits, honey and sesame (Wai, 2010). Some 62 per cent of the world's organic cotton is produced in South East Asia and another 28 per cent in the Middle East (Ferrigno, 2010). Sixty per cent of India's organic exports is bound for Europe and the rest to the USA, Japan and the Middle East (Wai, 2010). While China and India are mainly producing and exporting organic products, other countries in Asia such as Japan, South Korea, Taiwan, Singapore and Hong Kong are importing organic products from e.g. Europe, North America and Oceania (Willer and Kilcher, 2010).

Africa produces almost entirely for the export market with a large majority going to Europe (Bouagnimbeck, 2010). Most of the organic land in Africa is located in Uganda, Tunisia, Ethiopia, Tanzania, Sudan, South Africa and Egypt (Bouagnimbeck, 2010). Tunisia, Uganda and Egypt have the highest share of total land managed organically (1.1–1.8 per cent). The export of organic products covers a wide range: from vegetables (Egypt, etc.) and olives (Tunisia, etc.) produced in the North to coffee, cocoa, nuts, spices, cotton and dried fruits from the central part (Uganda, Tanzania, etc.) and fresh fruits from South Africa (Bouagnimbeck, 2010; Kledal and Kwai, 2010).

The environmental implications of the growing global trade with organic products, which is a concern for organic consumers, will be further discussed in the following section. The import of organic products from the South leads to some environmental impacts both at the farm and during transport. At the farm, the conversion to organic agriculture might have some environmental benefits compared to a former conventional production, while the long-distance transport from the farm to the consumers mainly has negative environmental consequences. The following two sections focus on the possible environmental benefits at the farm scale (comparing organic and conventional) and the extent of environmental impacts during long-distance transport.

Environmental impact of organic and conventional foods at the farm gate and local scale

The environmental impact of a product appearing at the retailer represents the aggregated impact from the production stage (input and farming) and the following treatment and transport. For most food products and most impact categories (but not all) the agricultural phase dominates the overall environmental impact. Thus, it is important to consider the impacts at the farm scale.

So what are the environmental implications of organic as compared to conventional production at the farm scale? Figure 3.3 gives a qualitative overview of environmental impacts of organic farming relative to conventional farming at the farming stage on the basis of a literature review by Schader et al. (2011) covering a range of impacts. Figure 3.3 shows an overall picture. Many of the studies behind the results were carried out in the industrialized world, but there is no reason to believe that the picture is not valid for developing and transition countries as well. The dots represent the result most frequently found in the literature. However, due to a high variability among supply chains, regions and products and due to methodological gaps in measuring impacts and comparing farming systems, it must be acknowledged that there is a range of uncertainty regarding some impact categories.

Impacts of organic farming on biodiversity range from much better to equal compared to non-organic agriculture. Due to lacking scientific evidence, Schader et al. (2011) conclude that both systems perform equally well in terms of genetic diversity. According to most studies, organic agriculture clearly performs better for faunal and floral species diversity (Bengtsson et al., 2005; Hole et al., 2005). Several authors also found a higher concentration of landscape elements (Norton et al., 2009; Schader et al., 2010) on organic farms.

Also, organic farming performs better in terms of less ground and surface water pollution, promoting air quality and soil fertility. Compared to conventional farming, water consumption for agriculture is not substantially affected by organic farming systems.

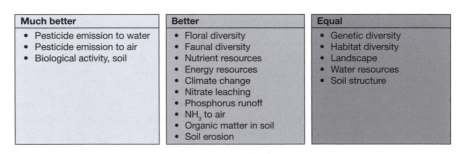

Much better	Better	Equal
• Pesticide emission to water • Pesticide emission to air • Biological activity, soil	• Floral diversity • Faunal diversity • Nutrient resources • Energy resources • Climate change • Nitrate leaching • Phosphorus runoff • NH_3 to air • Organic matter in soil • Soil erosion	• Genetic diversity • Habitat diversity • Landscape • Water resources • Soil structure

Figure 3.3 Environmental performance of organic farming compared to conventional farming at the farming stage

Source: After Schader et al., 2011.

Natural resources are in general positively affected by organic farming, though when considered per amount of product produced the picture is less clear. Nutrient resources are less depleted, e.g. due to less use of phosphorus fertilizers, fossil energy resources are spared due to higher energy efficiency for producing foodstuffs. As will be discussed in more detail below, there is contradictory evidence regarding climate change. Regarding N_2O emissions per kg of product evidence is most uncertain. The complexity in evaluating resource use arises because different impacts may occur in different stages of the production to a different extent, and thus trade-offs exist in resources used at different places in the chain, e.g. use of fertilizer in conventional production requires energy and thus increases energy consumption, but at the same time the yield of agricultural products is increased. The net result may be very little energy requirement per unit produced.

For consumer choice it is often assumed that it is the sum of impacts of a particular food that relates to the question: what environmental load am I putting on the globe by consuming this product compared to another product that fulfils more or less the same need? In this respect, life cycle assessment (LCA) is an internationally accepted method to assess environmental impacts along the life cycle of a product (Finnveden et al., 2009). This is true for a number of – but not all – environmental dimensions. LCA is particularly valid when evaluating use of fossil energy and its impact on global warming, which in nature are global concerns. The methodology is also well developed for assessing water and air pollution (eutrophication, acidification), which are important impacts related to food production as well, though the damage of the impact may depend on local conditions. Contrary to these, impact effects on e.g. biodiversity or ecotoxicity due to pesticides are only rarely assessed in LCA due to methodological difficulties (e.g. Thomassen et al., 2008; Halberg et al., 2010). Furthermore, the implementation of socio-economic impacts in LCA is still in its infancy (Griesshammer et al., 2006; Jørgensen et al., 2008).

It is at the farming stage that the differences between organic and conventional production/chains are most pronounced. Therefore, it is important to understand differences in environmental impact of organic and conventional foods at the farm gate.

Figure 3.4 presents the results of a review of LCA studies comparing greenhouse gas (GHG) emissions for organic and conventional products at the farm gate, thus including all inputs used at the farm and the emissions taking place at the farm. Most studies were carried out in a European or North American context, but a few studies are included from other parts of the world – oranges in Brazil and soya beans in China. Though LCA is a highly standardized methodology, the outcome of the assessment still relies on a number of assumptions and methodological choices, and these may highly impact the outcome of the assessment. Therefore, the most reliable assessment occurs when comparisons are performed within the same study, which was the case in the data presented in Figure 3.4.

Figure 3.4 indicates that overall there seems to be no major general differences between the GHG embedded in organic and conventional products, with 20 studies above the line, where organic performs better than conventional and 8 studies below the line where organic performs worse. The global warming impact of a product is essentially a result of N_2O emissions related to the circulating nitrogen in the agricultural system and nitrogen leaching, and the CO_2 emission related to energy use, e.g. for traction and production of fertilizer.

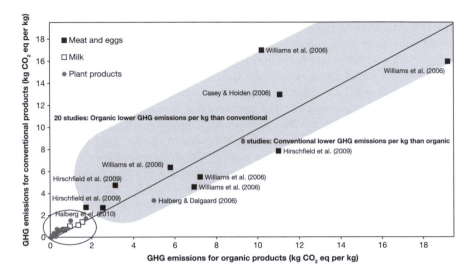

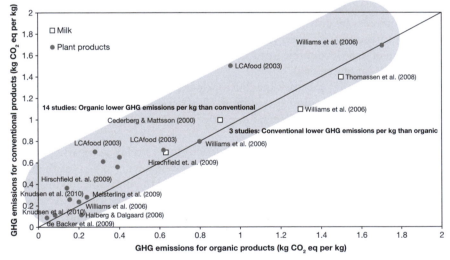

Figure 3.4 Greenhouse gas emissions per kg conventional product plotted against the organic counterpart. Organic perform better above the line and conventional better below the line. Upper graph contains the total number of LCA studies, whereas the lower graph is a zoom of the studies of milk and plant products.

Source: Knudsen, 2011.

For livestock products – and in particular products from ruminants – a major additional contributor is methane related to enteric fermentation and manure handling. The 'no effect' of organic practice from an overall point of view is related to the fact that circulating nitrogen in the systems is not that much different, in particular not in livestock systems, and in livestock systems the organic practice does not impact on methane emissions generally. A major difference is in the use of energy for fertilizer production, but at the same time yield per ha is often lower in the organic system, and when these effects are balanced out, no major difference exists.

When focusing on plant products only (lower part of Figure 3.4) 14 out of 16 studies, however, showed a lower or similar greenhouse gas emission for organic products compared to conventional. This is mainly caused by the avoidance of mineral fertilizer and pesticides, which in most cases are not outweighed in the CO_2 accounting by the lower or similar yields of organic production.

Apart from the factors already mentioned, the production system may impact on soil carbon changes and – in turn – the emissions of carbon dioxide to the atmosphere (impacting on global warming). This aspect, the soil carbon sequestration, was generally not included in the LCAs presented in Figure 3.4 due to methodological limitations. However, organic farming systems on average have a higher content of organic matter in the soil (e.g. Mäder et al., 2002; Mondelaers et al., 2009), and studies by Halberg et al. (2010) and Knudsen et al. (2010) indicate that the inclusion of estimated soil carbon changes in fact impacts on the global warming potential (GWP) of the products in favour of organic products. Knudsen (2011) showed that the difference in GWP for organic and conventional soybean was further increased by approx. 130 kg CO_2 eq. per ton of soybean, when including soil carbon changes for soybeans. This impact from including soil carbon changes is noteworthy compared to an initial carbon footprint of soybeans around 200 kg CO_2 eq. per ton of soybean (Knudsen et al. 2010). When including soil carbon changes in the carbon footprint of animal-based products, the relative impact is less pronounced due to a higher initial carbon footprint of animal-based products (Halberg et al., 2010). However, there is a need for further methodological development on how to estimate and include soil carbon changes in LCAs.

What has been presented until now is the global picture. However, within organic and conventional production systems a considerable variation may exist due to management decisions. Thus, Kristensen et al. (2011) showed a variation of more than 50 per cent in the GWP per kg for milk from conventional farms and of 25 per cent within organic Danish dairy farms. Similarly, Hvid et al. (2005) showed a variation in pork meat from different conventional farms of 37 per cent. Thus, within systems there is considerable room for improvement regarding environmental impact. Interestingly, Knudsen et al. (2011) showed that the GWP of oranges at the farm gate for smallholder organic farms in Brazil was 25 per cent lower than from a large-scale organic farm which in turn was similar to conventional small-scale farms. A similar pattern was found for contributions to eutrophication and acidification. Obviously, this has nothing to do with the scale of production, but instead the rationale for production. In this case, the large orange unit based its production on input substitution and used approximately twice as much manure N compared to the smallholders, which increased for example the estimated nitrous oxide emissions (Knudsen et al., 2011). Similar results were found regarding organic and conventional pear production in China (Liu et al., 2010).

Thus, there are reasons to believe that merely stating that a product is organically produced does not imply that the environmental impact expressed by global warming potential, eutrophication and acidification is lower than its conventional counterpart. However, as is also illustrated, organic products probably have the potential to perform better from an environmental point of view if sufficient care is taken in the rationale of the production, and other important impacts must be considered as well as is further discussed later. See also Case study 1 for an illustrative example of the climate change mitigation potential of organic agriculture.

Transport and its interaction with seasonal production

The transport of the food we eat, often referred to as food miles, has been widely debated during the last 10 to 15 years, especially in the UK (e.g. Kemp et al., 2010). Within the organic sector, this issue of 'food miles' has also been widely discussed especially with regard to the dilemma of airfreighted agricultural products from developing countries (Soil Association, 2007), which has also given rise to the term 'fair miles' (MacGregor and Vorley, 2006) referring to the aim of development and fair trade with regard to developing countries (Edwards-Jones et al., 2009).

The environmental impact of food transport is mainly related to emissions from energy use. That is in particular greenhouse gas emissions, but also particulate matter and acidification may be important depending on the type of fuel used. Nevertheless, GHG emissions are a good proxy for the overall effects, and the focus will again be on this in the following section.

Oxfam (2009) estimated for the UK situation that transport accounted for 12 per cent of greenhouse gas emissions from food consumption (Figure 3.5), so this aspect should not be ignored. It appears further that both domestic road freight and international road freight were major contributors.

The issue of transport can be distinguished between direct transport of food such as vegetables or indirect transport such as where imported feed supports a local organic livestock production.

Direct transport of food

The complexity of this issue is illustrated in a recent work by Knudsen et al. (2011). They investigated what processes contributed most to the GWP of organic orange juice produced in Brazil and consumed in Denmark. The

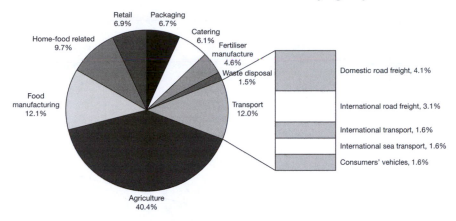

Figure 3.5 Greenhouse gas emissions from UK food consumption

Source: Oxfam, 2009.

processes included production at the farm, processing to frozen concentrated orange juice, transport of frozen concentrated orange juice to Germany where it was reconstituted, and transport of juice to Denmark. The results are shown in Case study 2 which follows this chapter.

Impacts from farm and processing are important, but more than 50 per cent of the total impact was related to transport. However, having a closer look at the transport, it appears that the main contributor (115/244) was related to transport of the reconstituted juice within Europe, whereas the intercontinental transport by ship was a minor contributor. Reasons are two. First, only the frozen concentrated orange juice (less weight) was transported a long way. Second, the long transport took place by ship. Thus, the impacts during transport depend on the distance, but even more on the transport mode. This is illustrated in Figure 3.6 which shows the greenhouse gas emissions per ton food transported by different means. Air freight has a much higher impact per ton km than sea freight.

From these data it can be estimated that the import for example of organic apples to Denmark from either Italy (by a 40-ton truck from Rome to a retail distribution centre (RDC) in Aarhus) or from Argentina (1,000 km by train within Argentina, 11,718 km by ship from Buenos Aires to Rotterdam and 834 km by 40-ton truck to RDC in Aarhus) would amount to:

Apples, Italy (Rome): $2,053$ km $\times$ 150 g CO_2 eq. per tkm = 308 kg CO_2 eq. per t product

Apples, Argentina (Buenos Aires): $(1,000$ km $\times$ 40 g CO_2 eq. per tkm$)$
$(11,718$ km $\times$ 9 g CO_2 eq. per tkm$)$ +
$(834$ km $\times$ 150 g CO_2 eq. per tkm$)$ =
271 kg CO_2 eq. per t product

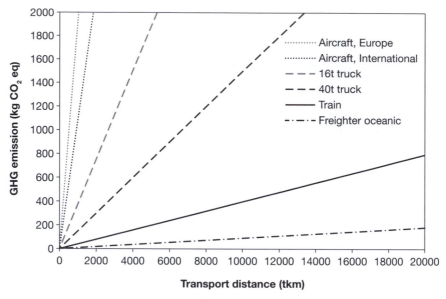

Figure 3.6 Greenhouse gas emissions per ton food transported as affected by transport
mode and distance

Source: Ecoinvent Centre, 2007.

This simple calculation indicates that the greenhouse gas emissions associated
with transport from Italy by truck is at the same level as from Argentina by ship.
In the example of apples from Italy the GHG contribution is around 0.3 kg
CO_2 eq. per kg. Comparing this number with the GWP from apple production per se (0.1 kg CO_2 eq. per kg (e.g Milà i Canals, 2006), transport
contributes very significantly in relative terms. On the other hand, if the food
in question were beef transported over the same distance, the contribution in
actual terms would be similar, but relatively the contribution from transport
would be approximately 1–2 per cent, since beef has so much higher GWP
(Figure 3.4).

Figure 3.7 presents the estimated actual transport contribution to the GHG
emissions of organic agricultural products that are imported to Denmark, based
on transport mode and distance. It is assumed that they are transported from the
capital city or the centre of the country of origin to a retail distribution centre
in Aarhus. Transport from Europe is assumed to be by road in 40-ton trucks,
whereas the products from the remaining countries are assumed to be transported by sea and reloaded in Rotterdam harbour to 40-ton trucks. It is assumed
that the organic agricultural products have the same required conditions during
transport and additional energy for cooled or refrigerated transport is not
included.

It is apparent that the estimated GHG emission related to transport does not
vary considerably depending on whether the organic products are transported

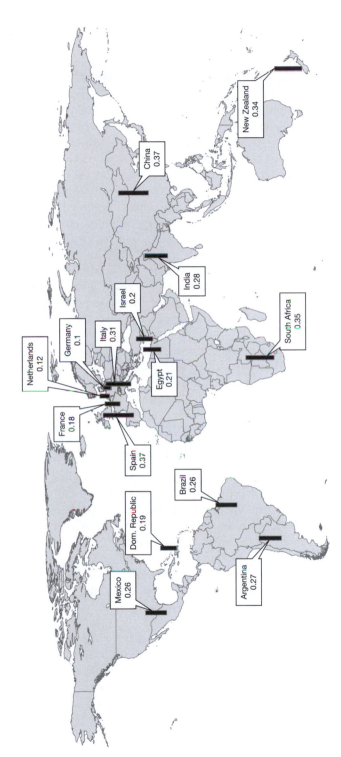

Figure 3.7 Estimated contribution from transport (kg CO$_2$ eq. per kg product) when importing organic products to Denmark from selected countries

Source: Knudsen, 2011.

from Southern Europe by truck or transported by ship from South America, Africa or Asia.

Some of the imported organic products can also be produced domestically, such as apples, tomatoes, salad, etc. The choice of whether to choose the domestically produced or imported products depends on the greenhouse gas emissions related to the products concerned during production, storage and transport. As a general rule domestic outdoor-produced vegetable foods usually have lower greenhouse gas emissions per unit when available at the supermarket in season, than imported. On the other hand imported produce may have lower environmental impacts if the domestic production uses heated or lit glasshouses or if the domestic produce is stored cold several months out of the domestic season (Milà i Canals et al., 2008). For apples consumed in Germany, Blanke and Burdick (2005) showed that the energy requirement of imported apples from New Zealand is 27 per cent higher compared to German apples stored at 1°C for five months. Milà i Canals et al. (2007b) also studied domestic versus imported apples and concluded that there are similarities in primary energy use of European and New Zealand apples during European spring and summer, while imported apples have a higher energy requirement during European autumn and winter, when local apples carry no CO_2 burden from cool storage.

Translating to the question of imported organic food or local conventional food one might conclude that for seasonal vegetable products probably the local foods most often will carry the lowest burden in season regarding GHG, while out of season the organic (transported) products often will carry no higher GHG load despite the transport.

Transport in relation to organic livestock production

While organic ruminant production to a wide extent relies on local feed resources, since grazing and other types of roughage are major components of feed, the same is not necessarily the case in pig and poultry production. While these species have to have access to roughage, the contribution of these feeds to the nutritional requirement of the animals is most often less than 10 per cent. Instead these species are fed with mixtures of cereals and seeds from oilseeds like soya bean and rape seed or legumes like peas. A large proportion of these foods can be imported to the farm from the local area or from long distance. Thus, while the livestock product may be produced locally, these products may still carry a significant indirect transport load.

Typically in pig and poultry production the feed input accounts for 60–70 per cent of the total greenhouse gas emissions in the production phase. The greenhouse gas emissions related to pork and eggs is typically 4–5 kg CO_2 eq. per kg product at the supermarket of which half is related to feed consumed (~2.5 kg CO_2 eq.) (Mogensen et al., 2009). Assuming that a third of the feed use is long-distance transported, the transport burden corresponds to 10 per cent of the total greenhouse gas emissions of the product.

Table 3.1 Environmental impact at farm for organic and conventional soybean produced in Jilin Province of China, and the impact related to transport to Denmark, per kg product

	GWP g CO_2 eq.	*Acidification g SO_2 eq.*	*Eutrophication g NO_3 eq.*
Organic, farm gate	156	2.3	5.0
Conventional, farm gate	263	4.5	13.0
Transport, China–DK	217	5.3	3.2

Source: Knudsen et al., 2010.

In particular the protein feed to livestock reared in Northern countries, in which the climate conditions are not suitable for oilseed production, are often long-distance transported. Knudsen et al. (2010) investigated in more detail the environmental impact of organic and conventional soybean produced in China and the contribution from transport, when transported to Denmark (Table 3.1).

While the environmental impact of organically produced soybeans is lower than the conventional soybeans, it appears from Table 3.1 that the contribution from transport in relation to global warming and acidification clearly 'overrules' the impact of production systems. Thus, transport is an important issue also in livestock production and unless it is ensured that the main part of the feed for the livestock is grown more or less locally, it may be a better choice to transport the livestock product than the feed.

In conclusion, transport of foods is a matter of concern regarding environmental impact and in particular greenhouse gas emissions, acidification and fossil energy use. This is true for transport of food as such and for transport of feeds for livestock. Thus, there are good reasons to focus on local production.

However, the mode of transport is more important than the total distance, and basically intercontinental transport by ship does not have a larger impact than truck transport across Europe or across the USA. And in situations where organic production can be performed 'environmentally cheaper' in the South than in the North (e.g. for out-of-season products), and the intercontinental transport takes place by ship, there is no reason to disregard the food with the long-distance transport from an environmental point of view.

Environmental aspects other than typically covered in LCA

While LCA is recognized as the best tool for assessing the life cycle impacts of products, the methodology does also have some unresolved problems (Reap et al., 2008a, 2008b) related to uncertainties in emission factors and impact estimation. In this respect ecotoxicity related to pesticide use is particularly critical due to lack of knowledge of the effects of particular pesticides and modes of transport, etc. Therefore, although theoretically the impact can be modelled in LCA, this aspect is often omitted.

Furthermore, appropriate methodologies are lacking for important impact categories for agricultural products. This includes impacts on soil quality and biodiversity (Milà i Canals et al., 2007a) which are parameters where organic production often differs from conventional production. Thus, when comparing organic and conventional products these aspects must be considered separately. Often they are left out completely.

The commonly used impact categories, such as global warming, eutrophication and acidification, are all based on emissions from the system. However, for biodiversity and soil the impacts are not necessarily caused by emissions, but rather changes in land use within the systems, which is one of the complicating factors for those impact categories. Furthermore, more than one single indicator might be relevant for biodiversity (such as plant diversity, faunal diversity, connectivity, etc.) and soil quality (such as soil organic matter, soil structure, soil erosion, soil pollution with copper, etc.). These difficulties, however, must not lead to the omission of these impacts from agricultural production systems as illustrated in Figure 3.3.

Biodiversity

As set out by the Millennium Ecosystem Assessment (2005), there is universal concern over the current scale and rate of losses of biodiversity worldwide, and the effects that this is likely to have on ecosystems and ecosystem services. There is plenty of evidence that use of herbicides diminishes floral diversity of the fields where they are used, and also in neighbouring hedgerows, etc. (Kleijn and Snoeijing, 1997). Besides affecting botanical diversity this in turn affects the numbers of insect and birds in the area through affecting the food availability for these species. Moreover, below-ground biodiversity is affected (OECD, 2001). Scale, however, is a critical issue here as the impact of changes in farming practice needs to be considered at both the farm and landscape scale. At the farm scale, biodiversity is affected by day-to-day management decisions around the use of agrochemicals, tillage, crop species, grazing intensity, etc. At the landscape scale, issues such as habitat fragmentation and water management are important. There are also some trade-offs between in-field and landscape-scale biodiversity such that low in-field biodiversity can to an extent be compensated for if the landscape is complex enough (Roschewitz et al., 2005).

As we have discussed, there is sufficient scientific evidence that organic farming practice benefits biodiversity due to lack of use of pesticides, but the extent to which this happens depends on maintaining a high biodiversity also on a structural scale – crop diversification, field size and hedgerows, etc. (Bengtsson et al., 2005; Hole et al., 2005).

Furthermore, it is now becoming generally accepted that, even if the processes are highly complex, there is a general positive relationship between diversity and stability and also between diversity and productivity. Pollination is an essential ecosystem service, vital to the maintenance both of wild plant communities and agricultural productivity. While the full effect of the value of ecosystem services like animal-based pollination is only poorly understood, it is estimated to represent a value corresponding to 10 per cent of the world's food production (Gallai et al., 2009).

Currently in LCAs, biodiversity is often covered using the indicators land transformation and occupation as proxies. The assumption is that different types of land transformation and occupation have the same impact on biodiversity. This approach leads to results showing a worse performance of organic foodstuffs regarding biodiversity, since land use per kg of organic product most often is higher compared to conventional agriculture, at least when organic farming is compared to intensive conventional production systems in industrialized countries. In developing countries under low-input and often poor agro-ecological growing conditions and in mixed systems, the crop yield per unit of land in organic systems is often similar to conventional systems, and thus no systematic difference in request for land use is seen (see Chapter 2).

However there are other relevant ways of considering biodiversity. Koellner and Scholz (2008) suggest introducing habitat quality aspects within the different types of land occupation and transformation and currently there are several research groups working on LCA-compatible biodiversity indicators (Curran et al., 2010; Jeanneret et al., 2008; Schader et al., 2010). If this succeeds it may have a major influence on the judgement of whether a particular organic product carries a higher or lower environmental burden than a comparable conventional product.

Soil fertility

Sustaining soil fertility is an important indicator for assessing the environmental impacts of agricultural systems and food products. However, just like biodiversity, soil fertility can currently not be covered within LCAs. The land-use model by Milà i Canals et al. (2007a) is a major step forward. Other very detailed models of soil fertility have been developed for the inclusion in general LCAs as well (Oberholzer, 2006). However, models are focused on specific regions and ecosystems and do not allow for impacts related to products. Furthermore, the high degree of detail of these models makes them unsuitable for general use in LCAs, due to the great data requirements.

Pesticide use and human health

Pesticide use poses a risk to human health. In developing countries, this is mainly related to the use of the pesticides (the spraying process) and risks related to storage, etc. Yanggen et al. (2003) and Konradsen et al. (2003) reported thousands of deaths related to pesticide use in developing countries and also stated that numbers are considered to be very much under-reported due to lack of reporting to health authorities. Similarly, it was shown that workers at organic tea farms in Sri Lanka had lower blood values of fluorinated organic compounds than workers on conventional farms (Guruge et al., 2005).

The unacceptable exposure to pesticides in conventional farming is related to lack of proper education as well as equipment and one may ask if this documented problem was a problem of the past. However, recent work from Brazil (Almeida and Abreu, 2009), from China and Sri Lanka (Qiao et al., 2009) and from India (Panneerselvam et al., 2011) have shown that the avoidance of pesticide use was one of the main arguments for farmers to convert to organic farming. The work of Panneerselvam et al. (2011) covered three states of India producing rice, mixed vegetables and cotton, where either 'avoidance of pesti- cides' or 'health benefits' was mentioned as number one or two of the highest priority advantages of organic farming, and in two out of three states con- ventional farmers in the same areas perceived organic farmers to have better health than conventional farmers.

In a survey of organic farmers in Wanzai and Wuyuan in China and in Kandy, Sri Lanka (Qiao et al., 2009) 29 per cent of the Chinese families reported that a household member had needed medical attention because of pesticides. The Sri Lankan case showed that introduction of organic farming also had changed the attitudes of the neighbouring conventional farmers towards using fewer pesticides. These reports show that organic farming can have major health bene- fits for the farmers and workers involved in the production. Thus while these benefits may be less pronounced when comparing organic and conventional farming in Europe and North America, no doubt it is an important aspect when considering impacts of organic farming in BRICS and developing countries.

In the data presented we used the functional unit one kg of product. The rationale of this is that one kg of organic product is equivalent to one kg of conventional product. Schader et al. (2011) however stress that hereby important differences between organic and conventional products are neglected. It is likely that higher willingness to pay is an indicator for a superior product quality (at least perceived by consumers) or different functions of the food stuffs. Thus, these additional functions should be reflected either in terms of a more differentiated functional unit(s) or monetary functional units when comparing the same type of product of different quality.

Conclusion

Organic production has some obvious benefits compared to conventional production such as avoidance of pesticides, especially for farmers applying poten- tially toxic pesticides in developing countries, beneficial effects on biodiversity and increased soil carbon sequestration on average compared to conventional farming. However, the remarkable growth in the global markets for organic food and drinks during the last decade, from 2002 to 2012, also holds a risk of increasing the environmental load from long-distance transport and of pushing the organic food and farming systems towards the conventional farming model and thereby diminishing the environmental benefits of organic agriculture. Transport often accounts for approximately 35–75 per cent in the carbon foot- print of organic plant products imported to e.g. UK or Denmark from Southern

Europe by road or other continents by sea. The actual contribution from transport of animal products is the same, but relatively it only accounts for 1–20 per cent. This effect needs to be considered and balanced with the other environmental benefits of organic food production, but in general terms this drawback can be considered as similar in size as out-of-season production in the Northern countries. However, it is also evident that in some cases the demand for organic products in the metropolitan centres and the North pushes the organic production towards more specialized production schemes based more on input substitution than ecological intensification which in turn diminish the benefits of organic food production. There is a need to consider how this dilemma in the global trade can be counteracted, by schemes for benchmarking and improvement of the organic production or regulations.

References

Almeida, G.F. and de Abreu, L.S. (2009) 'Estratégias produtivas e aplicação de princípios da agroecologia: o caso dos agricultores familiares de base ecológica da cooperativa dos agropecuaristas solidários de Itápolis – COAGROSOL', *Revista de Economia Agrícola*, 56, 1, 37–53

Barrett, H.R., Browne, A.W., Harris, P.J.C. and Cadoret, K. (2002) 'Organic certification and the UK market: organic imports from developing countries', *Food Policy*, 27, 301–318

Bengtsson, J., Ahnström, J. and Weibull, A-C. (2005) 'The effects of organic agriculture on biodiversity and abundance: a meta-analysis', *Journal of Applied Ecology*, 42, 261–269

Blanc, J. (2009) 'Family farmers and major retail chains in the Brazilian organic sector: assessing new development pathways. A case study in a peri-urban district of São Paulo', *Journal of Rural Studies*, 25, 322–332

Blanke, M.M. and Burdick, B. (2005) 'Food (miles) for thought – energy balance for locally-grown versus imported apple fruit', *Environmental Science & Pollution Research*, 12, 3, 125–127

Bouagnimbeck, H. (2010) 'Organic farming in Africa', in H. Willer and L. Kilcher (eds), *The World of Organic Agriculture. Statistics and Emerging Trends 2010*, IFOAM and FiBL, Bonn and Frick, pp. 104–110

Casey, J.W. and Holden, N.M. (2006) 'Greenhouse gas emissions from conventional, agri-environmental scheme, and organic Irish suckler-beef', *Journal of Environmental Quality*, 35, 231–239

Cederberg, C. and Mattsson, B. (2000) 'Life cycle assessment of milk production – a comparison of conventional and organic farming', *Journal of Cleaner Production*, 8, 49–60

Curran, M., de Baan, L., De Schryver, A., van Zelm, R., Hellweg, S., Koellner, T., et al. (2010) 'Toward meaningful end points of biodiversity in life cycle assessment', *Environmental Science & Technology*, 374–390

De Backer, E., Aertsens, J., Vergucht, S. and Steurbaut, W. (2009) 'Assessing the ecological soundness of organic and conventional agriculture by means of life cycle assessment (LCA). A case study of leek production', *British Food Journal*, 111, 10, 1028–1061

Ecoinvent Centre (2007) 'Ecoinvent Database v.2.0'. Swiss Centre for Life Cycle Inventories. Online at: www.ecoinvent.org/

Edwards-Jones, G., Plassmann, K., York, E.H., Hounsome, B., Jones, D.L. and Milà i Canals, L. (2009) 'Vulnerability of exporting nations to the development of a carbon label in the United Kingdom', *Environmental Science & Policy*, 12, 4, 479–490

FAO (2001) 'World markets for organic fruit and vegetables – opportunities for developing countries in the production and export of organic horticultural products'. International Trade Centre, Technical Centre for Agricultural and Rural Cooperation, Food and Agriculture Organization of the United Nations, Rome. Online at www.fao.org/docrep/004/y1669e/y1669e00.htm#Contents

Ferrigno, S. (2010) 'Organic cotton production and fiber trade 2008/09: in the eye of the storm', in H. Willer and L. Kilcher (eds), *The World of Organic Agriculture. Statistics and Emerging Trends 2010*, IFOAM and FiBL, Bonn and Frick, pp. 67–70

Finnveden, G., Hauschild, M.Z., Ekvall, T., Guninée, J., Heijungs, R., Hellweg, S., et al. (2009) Review. 'Recent developments in life cycle assessment', *Journal of Environmental Management*, 91, 1–21

Gallai, N., Salles, J.-M., Settele, J. and Vaissiere, B.E. (2009) 'Economic valuation of the vulnerability of world agriculture confronted with pollinator decline', *Ecological Economics*, 69, 810–821

Garibay, S.V. and Ugas, R. (2010) 'Organic farming in Latin America and the Caribbean', in H. Willer and L. Kilcher (eds), *The World of Organic Agriculture. Statistics and Emerging Trends 2010*, IFOAM and FiBL, Bonn and Frick, pp. 160–172

Giovannucci, D. and Pierrot, J. (2010) 'Is coffee the most popular organic crop?', in H. Willer and L. Kilcher (eds), *The World of Organic Agriculture. Statistics and Emerging Trends 2010*, IFOAM and FiBL, Bonn and Frick, pp. 62–66

Griesshammer, R., Benoît, C., Dreyer, L.C, Flysjö, A., Manhart, A., Mazijn, B., et al. (2006) 'Feasibility study: integration of social aspects into LCA'. UNEP-SETAC Life Cycle Initiative, Paris

Guruge, K.S., Taniyasu, S., Yamashita, N., Wijeratna, S., Mohotti, K.M., Seneviratne, H.R. et al. (2005) 'Perfluorinated organic compounds in human blood serum and seminal plasma: a study of urban and rural tea worker populations in Sri Lanka', *Journal of Environmental Monitoring*, 7, 4, 371–377

Halberg, N. and Dalgaard, R. (2006) 'Miljøvurdering af konventionel og økologisk avl af grøntsager' [in Danish]. Environmental assessment of conventional and organic production of vegetables. Arbejdsrapport fra Miljøstyrelsen nr. 5. Online at: http://www2.mst.dk/common/Udgivramme/Frame.asp?http://www2.mst.dk/udgi v/Publikationer/2006/87-7614-960-9/html/helepubl.htm

Halberg, N., Hermansen, J.E., Kristensen, I.S., Eriksen, J., Tvedegaard, N. and Petersen, B.M. (2010) 'Impact of organic pig production systems on CO_2 emission, C sequestration and nitrate pollution', *Agronomy for Sustainable Development*, 30, 721–731

Hirschfeld, J., Weiß, J., Preidl, M. and Korbun, T. (2008) 'The impact of German agriculture on the climate', Schriftenreihe des IÖW 189/08, p. 42

Hole, D.G., Perkins, A.J., Wilson, J.D., Alexander, I.H., Grice, P.V. and Evans, A.D. (2005) 'Does organic farming benefit biodiversity?', *Biological Conservation*, 122, 113–130

Hvid, S.K., Nielsen, P.H., Halberg, N. and Dam, J. (2005) 'Miljøinformation i produktkæden – Et casestudie af produktkæde med svinekød', Miljøprojekt Nr. 1027, Danish Ministry of the Environment, Copenhagen

IFOAM (2005) 'Principles of Organic Agriculture'. International Federation of Organic Agriculture Movements (IFOAM). Online at: www.ifoam.org/organic_facts/ principles/pdfs/IFOAM_FS_Principles_forWebsite.pdf

Jeanneret, P., Baumgartner, D., Freiermuth-Knuchel, R. and Gaillard, G. (2008). 'A new LCIA method for assessing impacts of agricultural activities on biodiversity (SALCA-Biodiversity)', in T. Nemecek, and G. Gaillard (eds), *Proceedings of the 6th International Conference on Life Cycle Assessment in the Agri-Food Sector. Towards a sustainable management of the food-chain*, November 12–14, Zürich, Agroscope Reckenholz Tänikon (ART, Zurich), pp. 34–39

Jørgensen, A., Le Bocq, A. and Hauschild, M.Z. (2008) 'Methodologies for Social Life Cycle Assessment – a review', *International Journal of Life Cycle Assessment*, 13, 96–103

Kemp, K., Insch, A., Holdsworth, D.K. and Knight, J.G. (2010) 'Food miles: Do UK consumers actually care?' *Food Policy*, Advance Online Publication, http://dx.doi.org/10.1016/j.foodpol.2010.05.011

Kledal, P. (2009) 'The four food systems in developing countries and the challenges of modern supply chain inclusion for organic small-holders'. Paper for the International Rural Network Conference in India, Udaipur, 23–28 August

Kledal, P. and Kwai, N. (2010) 'Organic food and farming in Tanzania', in H. Willer and L. Kilcher (eds), *The World of Organic Agriculture. Statistics and Emerging Trends 2010*, IFOAM and FiBL, Bonn and Frick, pp. 111–115

Kleijn, D. and Snoeijing, G.I.J. (1997) 'Field boundary vegetation and the effects of agrochemical drift: botanical change caused by low levels of herbicide and fertilizer', *Journal of Applied Ecology*, 34, 1413–1425

Knudsen, M.T. (2011) 'Environmental assessment of imported organic products – focusing on orange juice from Brazil and soybeans from China', PhD thesis, Faculty of Life Sciences, University of Copenhagen and Faculty of Agricultural Sciences, Aarhus University, Denmark

Knudsen, M.T. and Halberg, N. (2007) 'How to include on-farm biodiversity in LCA in food', Contribution to the LCA Food Conference, April, Gothenburg, Sweden

Knudsen, M.T., Yu-Hui, Q., Yan, L. and Halberg, N. (2010) 'Environmental assessment of organic soybean (Glycine max.) imported from China to Denmark: a case study', *Journal of Cleaner Production*, 18, 1431–1439

Knudsen, M.T., de Almeida, G.F., Langer, V., de Abreu, L.S. and Halberg, N. (2011) 'Environmental assessment of organic juice imported to Denmark: a case study on oranges (Citrus sinensis) from Brazil', *Organic Agriculture*, 1, 167–185

Koellner, T. and Scholz, R. (2008). 'Assessment of land use impacts on the natural environment', *International Journal of Life Cycle Assessment*, 13, 1, 32–48

Konradsen, F., Van der Hoek, W., Cole, D.C., Hutchinson, G., Daisley, H., Singh, S. and Eddleston, M. (2003) 'Reducing acute poisoning in developing countries: options for restricting the availability of pesticides', *Toxicology*, 192, 249–261

Kristensen, T., Mogensen, L., Knudsen, M.T. and Hermansen, J.E. (2011) 'Effect of production system and farming strategy on greenhouse gas emissions from commercial dairy farms in a life cycle approach including effect of different allocation methods', *Livestock Science*, 140, 136–148

LCAfood (2003) Online at www.lcafood.dk

Liu, Y., Langer, V., Høgh-Jensen, H. and Egelyng, H. (2010) 'Life cycle assessment of fossil energy use and greenhouse gas emissions in Chinese pear production', *Journal of Cleaner Production*, 18, pp. 1423–1430. http://orgprints.org/18291/

MacGregor, J. and Vorley, B. (2006) 'Fair miles? The concept of 'food miles' through a sustainable development lens', International Institute for Environment and Development (EEED), London

Mäder, P., Fließbach, A., Dubios, D., Gunst, L., Fried, P. and Niggli, U. (2002), 'Soil fertility and biodiversity in organic farming', *Science*, 296, pp. 1694–1697

Meisterling, K., Samaras, C. and Schweizer, V. (2009) 'Decisions to reduce greenhouse gases from agriculture and product transport: LCA case study of organic and conventional wheat', *Journal of Cleaner Production*, 17, 222–230

Milà i Canals, L. (2006) 'Evaluation of the environmental impacts of apple production using life cycle assessment (LCA): case study in New Zealand', *Agriculture, Ecosystems and Environment*, 114, 226–238

Milà i Canals, L., Romanya, J. and Cowell, S. (2007a) 'Method for assessing impacts on life support functions (LSF) related to the use of "fertile land" in Life Cycle Assessment (LCA)', *Journal of Cleaner Production*, 15, 1426–1440

Milà i Canals, L., Cowell, S.J., Sim, S. and Basson, L. (2007b) 'Comparing domestic versus imported apples: a focus on energy use', *Environmental Science & Pollution Research*, 14, 5, 338–344

Milà i Canals, L., Muñoz, I., Hospido, A., Plassmann, K. and McLaren, S. (2008) 'Life Cycle Assessment (LCA) of domestic vs. imported vegetables. Case studies on broccoli, salad crops and green beans', CES Working Paper 01/08, University of Surrey, Guildford

Millennium Ecosystem Assessment (2005) 'Millennium Ecosystem Assessment', Online at www.maweb.org/en/index.aspx

Mogensen, L., Hermansen, J.E., Halberg, N., Dalgaard, R., Vis, J.C. and Smith, B.G. (2009) 'Life Cycle Assessment across the food supply chain', in C. Baldwin (ed,), *Sustainability in the Food Industry,* Wiley-Blackwell and the Institute of Food Technologists, USA, pp. 115–144

Mondelaers, K., Aertsens, J. and Van Huylenbroeck, G. (2009) 'A meta-analysis of the differences in environmental impacts between organic and conventional farming', *British Food Journal*, 111, 1098–1119

Norton, L., Johnson, P., Joys, A., Stuart, R., Chamberlain, D., Feber, R., et al. (2009) 'Consequences of organic and non-organic farming practices for field, farm and landscape complexity', *Agriculture, Ecosystems and Environment*, 129, 221–227

Oberholzer, H. (2006). 'Methode zur Beurteilung der Wirkungen landwirtschaftlicher Bewirtschaftung auf die Bodenqualität in Ökobilanzen', Zürich, ART

OECD (2001) *Valuation of Biodiversity Benefits Selected Studies*, Paris, OECD

Organic Monitor (2010) *The Global Market for Organic Food and Drink: Business Opportunities and Future Outlook* (3rd edition). www.organicmonitor.com/700340.htm

Oxfam (2009) '4-a-week – Changing food consumption in the UK to benefit people and planet', Oxfam GB Briefing Paper, March

Panneerselvam, P., Halberg, N., Vaarst, M. and Hermansen, J.E. (2011) 'Indian farmers' experience with and perceptions of organic farming', *Renewable Agriculture and Food Systems,* http://dx.doi.org/10.1017/S1742170511000238

Qiao, Y., Setboonsarng, S. and Halberg, N. (2009) 'PRC Country study on organic agriculture and the Millennium Development Goals', ADBI Working Paper, unpublished.

Reap, J., Roman, F., Diuncan, S. and Bras, B. (2008a) 'A survey of unresolved problems in life cycle assessment – Part 1: Goal and scope and inventory analysis', *International Journal of Life Cycle Assessment*, 13, 290–300

Reap, J., Roman, F., Diuncan, S. and Bras, B. (2008b) 'A survey of unresolved problems in life cycle assessment – Part 2: impact assessment and interpretation', *International Journal of Life Cycle Assessment*, 13, 374–388

Refsgaard, K., Jenssen, P.D. and Magid, J. (2006) 'Possibilities for closing the urban-rural cycles', in Halberg, N. Alosem H.F., Knudsen, M.T. and Kristensen E.S. (eds) *Global Development of Organic Agriculture: Challenges and Prospects*, Wallingford, CABI, pp. 181–213

Rigby, D. and Bown, S. (2003) 'Organic food and global trade: is the market delivering agricultural sustainability?', School of Economic Studies Discussion Paper, Manchester University, pp. 1–21 www.socialsciences.manchester.ac.uk/disciplines/economics/research/discussionpapers/pdf/Discussion_paper_0326.pdf

Roschewitz, I., Gabriel, D., Tscharntke, T. and Thies, C. (2005) 'The effects of landscape complexity on arable weed species diversity in organic and conventional farming', *Journal of Applied Ecology*, 42, 873–882

Schader, C., Stolze, M. and Gattinger, A. (2011) 'Environmental performance of organic agriculture', in Boye, J. and Arcand, Y. (eds), *Green Technologies in Food Production and Processing*, Springer, New York.

Schader, C., Drapela, T., Markut, T., Hörtenhuber, S., Lindenthal, T., Meier, M. and Pfiffner, L. (2010) 'Biodiversity impact assessment of Austrian organic and conventional. dairy products', LCA Discussion Forum: Integrating biodiversity in LCA, November, Lausanne, Switzerland

Sheng, J.P., Shen, L., Qiao, Y.H., Yu, M. and Fan, B. (2009) 'Market trends and accreditation systems for organic food in China', *Trends in Food Science & Technology*, 20, 9, 396–401

Soil Association (2007) 'Should the Soil Association tackle the environmental impact of air freight in its organic standards?' Soil Association Standards Consultation – Air Freight Green Paper

StatBank Denmark (2010) Statistics Denmark. Online at: www.statbank.dk

Thomassen, M.A., van Calker, K.J., Smits, M.C.J., Iepema, G.L. and de Boer, I.J.M. (2008) 'Life cycle assessment of conventional and organic milk production in the Netherlands', *Agricultural Systems*, 96, 95–107

Twarog, S. (2006) 'Organic agriculture: a trade and sustainable development opportunity for developing countries' (Chapter 3, Part I), in *Trade and Environment Review*, United Nations Conference on Trade and Development (UNCTAD), United Nations, Rome

Wai, O.K. (2010) 'Organic Asia 2010', in H. Willer and L. Kilcher (eds) *The World of Organic Agriculture. Statistics and Emerging Trends 2010*, IFOAM and FiBL, Bonn and Frick, pp. 122–134

Willer, H. and Kilcher, L. (2010) *The World of Organic Agriculture. Statistics and Emerging Trends 2010*, IFOAM, Bonn and FiBL, Frick

Williams, A.G., Audsley, E. and Sandars, D.L. (2006) 'Determining the environmental burdens and resource use in the production of agricultural and horticultural commodities', in Main Report, Defra Research Project IS0205, Cranfield University and Defra, http://www.defra.go.uk

Yanggen, D., Crissman, C. and Espinoza, P. (2003) Los Plaguicidas: Impactos en Producción, Salud y Medio Ambiente en Carchi, Ecuador. Centro Internacional de la Papa (CIP), Instituto Nacional Autónomo de Investigaciones Agropecuarias (INIAP), and Ediciones Abya-Yala, Quito, Ecuador

Yin, S., Wu, L., Dub, L. and Chena, M. (2010) 'Consumers' purchase intention of organic food in China', *Journal of the Science of Food and Agriculture*, 90, 1361–1367

Case study 2
Life cycle assessment of organic orange juice imported from Brazil to Denmark

Marie Trydeman Knudsen, Niels Halberg,
Gustavo Fonseca de Almeida,
Lucimar Santiago de Abreu, Vibeke Langer
and John E. Hermansen

This case study presents a life cycle assessment (LCA) of organic oranges produced by small-scale farms in Brazil and processed into organic orange juice sold in Denmark. For further detailed description, please see Knudsen et al. (2011) (http://orgprints.org/18417/). The primary aim of the study was to identify the environmental hotspots in the product chain of organic orange juice originating from Brazil and sold in Denmark. The secondary aim was to compare the environmental impacts in the production of organic oranges at small-scale farms with conventional small-scale farms and organic large-scale farms in a case study area in São Paulo, Brazil.

Orange juice: from Brazilian farms to Danish consumers

We surveyed a number of organic and conventional farms in São Paulo, Brazil, and followed the product chains to Denmark. Inputs, outputs and emissions were inventoried on the farms and in the processing and transport.

The organic oranges are produced at small-scale farms in the State of São Paulo, Brazil. After harvest, the oranges leave the farms and travel approximately 120 km to the first processing plant where they are turned into frozen concentrated orange juice (FCOJ). The frozen concentrate is then transported to the harbour of Santos and shipped to Rotterdam. Here, the FCOJ is reloaded into trucks, transported 530 km and reconstituted into orange juice in a factory in Germany. When the orange juice is packed, it travels almost 900 km to Denmark.

The hotspot analysis studied the orange juice in all stages until delivery in Denmark, including input production and farm stage (e.g. N_2O emissions) in Brazil and processing (e.g. concentration, reconstitution, packing) and transport (including transport between all the steps and to final destination in Denmark). The studied unit for the hotspot analysis was 'one litre of organic orange juice'.

Figure 1 Orange production in Brazil and barrels with frozen concentrated orange juice

In addition to the hotspot analysis, the environmental impacts from the orange production at small-scale organic farms were compared to small-scale conventional farms or large-scale organic farms. The studied unit for the comparison was 'one tonne of oranges leaving the farm gate' and the study therefore only focused on the emissions during input production and at the farm.

Life cycle assessment as a tool

We used life cycle assessment (LCA) as a tool to calculate e.g. how much transport matters, with regard to climate and environment, when organic orange juice is imported to Denmark. Life cycle assessment includes all the relevant environmental impacts from the product chain – from the production of inputs, over the emissions at the farm to the processing and the transport of the product to Denmark. Finally, the environmental impact per kg product can be calculated, which for global warming is expressed in g CO_2 equivalents per kg orange juice. CO_2 equivalents are the sum of the climate gases CO_2, N_2O and CH_4 where the gasses are weighted according to their effect on the climate in the atmosphere in relation to CO_2. Soil carbon sequestration was only included in the sensitivity analysis.

The impact categories considered were effects on climate change (global warming potential), eutrophication, acidification, non-renewable energy use and land use.

Results from the hotspot analysis

The total global warming potential of the orange juice from organic small-scale farms was 423 g CO_2 equivalents per kg orange juice. The hotspot analysis showed that the transport stage contributes 58 per cent to the total greenhouse gas emissions followed by the farm stage (mainly N_2O from the plantation and CO_2 from the traction) and the processing (mainly CO_2).

Interestingly, the main contribution to the transport stage came from the truck transport of the juice and the truck transport of oranges from the farm to the processing plant, where 'water' is transported. The long-distance ship transport of the frozen concentrated orange juice only contributed 6 per cent to the

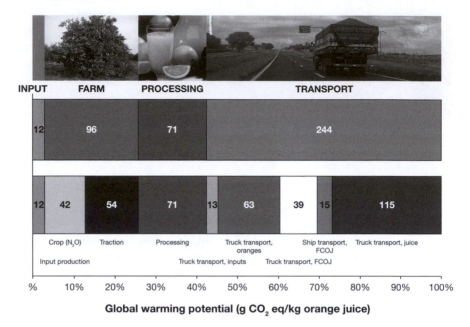

Figure 2 Greenhouse gas emissions from organic orange juice made from oranges
produced by small-scale organic farms in the state of São Paulo, Brazil,
processed into orange juice from concentrate and transported to Denmark: a
hotspot analysis

transport stage, since the orange is concentrated to approximately one-tenth of
the original weight.

Comparison of organic and conventional orange production

In the study, small-scale organic and conventional orange plantations were
compared. The main differences were higher crop diversity and the replacement
of toxic pesticides with small amounts of $CuSO_4$ and $CaSO_4$ at the small-scale
organic farms. These factors may have a positive impact on biodiversity and on
the health of the farmers applying the pesticides. Though non-significant,
nominally the environmental impacts per ton organic oranges were on average
65 per cent of the conventional, except from the land use (Figure 3). On a per
hectare basis, the organic small-scale plantation had a significantly lower non-
renewable energy use and acidification potential compared to the small-scale
conventional plantations (Figure 4).

Comparison of small-scale and large-scale organic orange production

Furthermore, the small-scale organic farms were compared to the large-scale organic orange plantations. The results showed that the large-scale organic plantations had significantly higher greenhouse gas emissions, eutrophication potential (Figure 4), copper use (towards plant diseases), and lower crop diversity per hectare compared to the small-scale organic farms. The same tendency was visible when the environmental impacts were accounted per kg oranges (Figure 3). These observations suggest that there is a need for more focus on how different certified organic productions comply with the organic ideas and principles. There might also be a need to scrutinize the organic regulations with regard to regulating especially greenhouse gas emissions and biodiversity.

A sensitivity analysis showed that inclusion of soil carbon changes in the life cycle assessment decreased the global warming potential of organic oranges from small-scale and large-scale producers to 51 and 88 kg CO_2 eq. per ton oranges, respectively, increasing the difference in greenhouse gas emissions per tonne organic and conventional oranges.

Conclusion

The transport stage had a major contribution (58 per cent) on the greenhouse gas emissions from the organic orange juice imported to Denmark, followed by

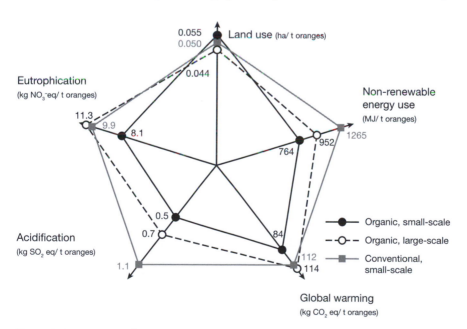

Figure 3 Environmental impacts per ton organic and conventional oranges produced in the state of São Paulo, Brazil

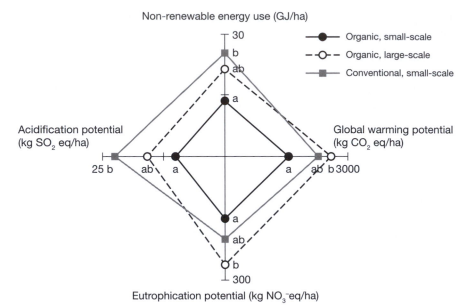

Figure 4 Environmental impacts per hectare organic and conventional oranges
produced in the state of São Paulo, Brazil

the farm stage (23 per cent). However, the major contribution to the transport
stage did not come from the long-distance sea transport, but from the truck
transport of fresh oranges in Brazil and ready-to-drink orange juice in Europe
where 'water' is transported.

Comparing the environmental impact from organic and conventional oranges
from small-scale farms, the main difference was higher crop diversity, replacement of toxic pesticides with small amounts of $CuSO_4$ and $CaSO_4$, and lower
non-renewable energy use per hectare.

Comparing the environmental impacts from organic oranges from small-scale
and large-scale farms, crop diversity was higher on the small-scale farms, while
global warming potential, eutrophication potential and copper use per hectare
were significantly lower.

Including soil carbon changes in the life cycle assessment widened the difference in greenhouse gas emissions between organic and conventional oranges.

There might be a need for assessing how different certified organic productions comply with the organic principles especially with regard to greenhouse
gas emissions and biodiversity.

4 The use of agro-ecological methods in organic farming[1]

Yuhui Qiao, Niels Halberg and Myles Oelofse

Introduction

The principles of organic agriculture express the core idea that agriculture and farming should emulate and sustain living ecological systems and cycles. Furthermore, organic agriculture should sustain and enhance the health of soil, plant, animal and planet and be managed in a precautionary and responsible manner to protect the health and well-being of current and future generations and the environment (EU and IFOAM Principles of Organic Agriculture). This should be achieved mainly by the appropriate 'design and management of biological processes based on ecological systems using natural resources, which are internal to the system' (IFOAM, 2005).

Several case studies describe successful and innovative implementation of these principles in terms of so-called agro-ecological methods. Agro-ecological methods include practices such as the recycling of nutrients through different composting methods, and use of legumes and multipurpose trees; pest regulation through intercropping in time or space and crop rotations; diversification via agro-forestry or integration of fish or livestock in cash crop farms (Pimentel et al., 2005; Mäder et al., 2002).

However, little is known regarding to what extent organic farmers in developed and developing countries actually use agro-ecological methods. Some studies suggest that certain types of organic farms in reality employ few such methods and thus do not actively seek to follow the principles. Darnhofer et al. (2010) reviewed the use of the term 'conventionalization' to describe a certain (development) type of organic farm, which besides following the certification requirements do little else to practise the organic principles and mainly replace chemical fertilizer with imported manure from intensive livestock farms.

If one purpose of promoting organic agriculture is to support an alternative development pathway for food production, which is based on organic principles and a balance between food production and other eco-system services, it may be a critical feature of certified organic production that the farms are actually developing and adopting agro-ecological methods. Thus, a relevant question is: Are some farms 'more organic' than others? This question is also relevant from a livelihood perspective since the idea that low-input and organic agriculture may benefit smallholder farmers is that they may replace cash-demanding external inputs with agro-ecological methods, rather than shift to a reliance on alternative, permitted external inputs.

The purpose of this chapter is to present innovative examples of the idea of using agro-ecological methods in organic agriculture; to give a critical assessment of the actual abundance of such methods in organic agriculture and to suggest ways to distinguish between different organic farming systems and the degree to which they actually employ organic principles in the farming practices.

Organic agriculture principles and agro-ecological measures ideally applied

The IFOAM Basic Standards for Organic Agriculture Production and Processing (IFOAM, 2005) include principles and recommendations for 'organic eco-systems' where provisions are made to 'maintain a significant portion of farms to facilitate biodiversity and nature conservation', including (among others) wildlife refuge habitats and wildlife corridors that provide linkages and connectivity to native habitat.

Central elements of the organic principles, which promote agro-ecological approaches, focus on farming practices that are: based on natural and local conditions; make full use of intrinsic resources of agro-ecosystems; reduce external inputs; diversify the farming systems and crops; and integrate management of agro-ecosystems, as listed in Figure 4.1.

Agro-ecological approaches are adopted in agricultural management in order to achieve the goals of sustainable agriculture. Figure 4.1 shows how the goals of sustainable agriculture are inextricably linked to agro-ecological measures.

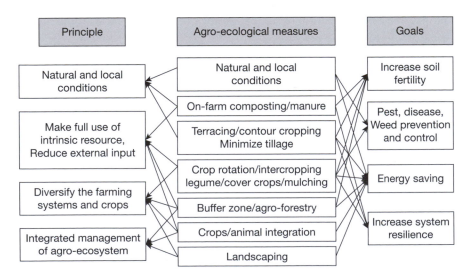

Figure 4.1 Principles of organic farming and agro-ecological measures

Organic farming relies on ecological processes, biodiversity and cycles adapted to local conditions, rather than the use of inputs with adverse effects: 'Organic agriculture combines tradition, innovation and science to benefit the shared environment and promote fair relationships and a good quality of life for all involved' (IFOAM, 2005). The overall goal is to achieve and maintain sufficient yields and quality for food security and farm income.

Soil fertility is a critical issue in organic agro-ecosystems especially for carbon (C) and nitrogen (N) management (Stockdale et al., 2001). Besides crop rotation and intercropping, the main agro-ecological measures include treating the soil with farmyard manure/composting; crop residue return/straw residue return after feeding; also application with green manure, legume crops and cover crops/mulching. These are integrated with plant and animal production; the recycling of wastes and by-products of plant and animal origin as input in plant and livestock production. Soil conservation is the basis of the whole organic farming system. Soil and water conservation can be achieved through agro-forestry, mulching, compost, terracing, contour cropping, minimizing tillage practice.

Pest, disease and weed management in organic agriculture is ideally based on a holistic approach, where preventive and control measures are integrated into system design. Management should thus promote environmental conditions that limit the propagation of diseases and pests and growth of weeds, but are favourable to the multiplication of natural enemies with the aim of maintaining the balance and biodiversity of agricultural ecosystems, and mitigation of the losses from various disease, pest and weed attacks. Priority should be given to promoting preventive measures including using resistant crop genotypes suitable to local conditions, crop rotation, intercropping, cover crops or mulching as well as buffer zones or agro-forestry. Pest and disease biological control can also be obtained through use of lamps, colour traps, pheromone traps, mechanical traps, etc. or by using locally available materials and authorized biological treatments such as neem, garlic, ashes, pyrethrum, etc.

Diversity means creating a heterogeneous environment both inside and out-side of cultivated fields and organic farms to increase the stability of the eco-system and its function such as productivity. Biological diversity in agricultural systems (agro-biodiversity), ranging from the genetic to system level, is considered a prerequisite for the development of sustainable agricultural systems (Love and Spaner, 2007). The promotion, and maintenance, of agro-biodiversity is also considered one of the key elements of organic agriculture (Parrot and Marsden, 2002). Intercropping is also one of the pillars of traditional agriculture, which organic farming confirms in all its potential and versatility in adapting to environmental and cultural contexts with different styles: mixed intercropping, intercropping in rows, strip intercropping, etc. Agro-forestry and planting hedgerows as buffer zones can enhance the biodiversity inside and outside the fields. Plant production can be integrated with animal production as a more diverse system.

Fostering agro-biodiversity by using agro-ecological methods can potentially improve the resilience of farming systems and create synergy between food security and other ecosystem services (Badgley et al., 2007; Perfecto and Vandermeer, 2008; see also Chapter 5 on climate change adaptation). Organic farming should protect and benefit biodiversity (e.g. rare species), air and water and the role of biodiversity in resilience. Those who produce organic products should protect and benefit the common environment including landscapes and habitats.

Organic agriculture should avoid the use of non-renewable resources and off-farm inputs that may have negative ecological and health effects. Inputs should be reduced by recycling and efficient management of materials and energy on the farm in order to maintain and improve environmental quality and conserve resources. External inputs like fertilizer and pesticides, etc. can be minimized through the appropriate management practices and methods which can save energy.

There is a large body of literature dealing with prescriptive practices, what types of management practices organic farmers should adopt, focusing upon crop rotations, nutrient management and pest, disease and weed management. A number of studies of organic cash crop production in developing countries have documented potential economic benefits for smallholder farmers (Lyngbæk et al., 2001; IFAD, 2005). However, the documentation of the use of organic principles and agro-ecological methods is scarce (Eyhorn, 2007; Parrot et al., 2006; Altieri, 2002). Literature presenting the effects of conversion on farmers' practices and to what degree agro-ecological practices are adopted practically in organic agriculture is rather limited. This will be discussed in the third section of this chapter.

Agro-ecological practices adopted practically in organic agriculture: status, characteristics and case studies

Under most circumstances, change is an essential process when farmers convert to organic agriculture. An agro-ecological approach, defined by Altieri (1995) as the design and management of agro-ecosystems applying ecological concepts, stresses system redesign as a principle. Organic agriculture should attain ecological balance and resilience (see above) through the design of cropping/farming systems. Most organic operations will reflect all of these to a greater or lesser degree. Since each farm is a distinct entity, there is a large degree of variation.

The type of changes that might occur when adopting organic agriculture and the degree to which these changes are of an agro-ecological nature depend upon a large range of factors, particularly the typology of the farm prior to conversion.

Status and characteristics of adopting agro-ecological methods in organic agriculture

When discussing farmers' adoption of agro-ecological practices, it is imperative to differentiate between certified and non-certified organic farmers. Certified farms operate within a legal framework according to organic standards and principles and are subject to inspections that conform to varying extents to agro-ecological principles. Our discussion mainly focuses on this type of organic agriculture.

Farmer adoption of agro–ecological practices following conversion to organic agriculture depends very much upon the type of system prior to conversion, which in turn has a strong bearing on the extent of changes occurring on-farm following adoption. Two broad and rather generalized typologies can be delineated. The first is traditional farms, characterized by a high degree of diversity and a low reliance on external inputs. The second is high-input farms, characterized by a low degree of diversity and a high reliance upon external inputs.

In developing countries, organic farming systems often take as a starting point concepts that have already been applied by traditional systems such as mixed cropping with the aim of improving the applied techniques. In some cases, traditional agricultural systems with low external input provide a potential basis for organic agriculture as a development option. But it is important to distinguish low-yielding low-input systems from modern organic agriculture as a development model. The principles of organic farming support an idea of eco-functional intensification, building practices that deliberately integrate traditional farming methods and make use of locally available resources.

Organic systems employing more agro-ecological practices

Conversion to organic agriculture typically means that farmers, first, need to rethink their management strategy for nutrient supply as well as pest, weed and disease management. The two main options for farmers are adjusting crop rotation to include fertility-building crops, and input substitution.

A review of the impact of organic agriculture in Latin America, conducted by IFAD, shows that the adoption of organic farming led to an increase in implementation of fertility-building crop rotations, the introduction of soil-conservation measures (e.g. terraces and live barriers), the implementation of new management practices such as agro-forestry and increased conservation of natural forests leading to greater biodiversity (IFAD, 2002). A review of organic agriculture in an Asian context by IFAD (2005) shows that the adoption of organic agriculture posed a challenge to 'intensive farmers' as the required changes were considered radical. Some forms of organic farming actively promote crop diversification like multistrata organic coffee farms in Costa Rica (Lyngbæk et al., 2001). Organic agriculture has a high and possibly decisive potential for reversing the dramatic decline of biodiversity (Scialabba and Hattam, 2002; Parrot et al., 2006).

A case study in Wanzai, China (Qiao et al., 2009; see also Case study 3, which follows this chapter), indicated a clear difference in the use of agro-ecological methods such as composting, intercropping and insect traps between the organic and conventional farms and also in the diversity of crops grown and livestock reared (Table 4.1). Moreover, the efforts to maintain soil fertility through the use of legumes as intercrops and other soil conservation methods (mulching) were significantly more widespread in the organic villages in Wanzai. It is evident when visiting the organic villages that widespread experimentation in methods for biological pest control and soil fertility improvement, etc. is taking place, which again presumably will benefit overall biodiversity and environmental preservation (Qiao et al., 2009).

Data from developed countries also prove this characteristic of employing more agro-ecological practices in organic systems. Compared with 15 conventional farms, 18 organic farms in the area of Tuscia, Italy, were also found to have a lower utilized agricultural area, higher woodland area, smaller plots with higher margin structure, more widespread leguminous crops for soil fertility resilience and more animals, with a higher integration of plant production and husbandry (Caporali, 2004). Mäder et al. (2002) from FiBL Switzerland reported results from a 21-year study of agronomic and ecological performance of biodynamic, bio-organic and conventional farming systems in Central Europe and found that organically manured, legume-based crop rotations utilizing organic fertilizers from the farm itself are a realistic alternative to conventional farming systems; they enhanced soil fertility and higher biodiversity in organic plots to render these systems less dependent on external inputs.

Long transition from input substitution to system redesign

In certified organic agriculture, the first changes typically entail fulfilling the requirements stipulated by the organic regulations. The on-farm changes brought about will depend upon whether the farm under conversion was previously high input or traditional. Farms with a previous high reliance upon external inputs for the management of soil fertility and pests, disease and weeds typically, as a first step, have a continued high reliance upon the import of

Table 4.1 Use of agro-ecological methods and planned diversity in organic and conventional households in Wanzai, China

Agro-ecological methods	Total sample	OA farmers	CA farmers	P
Number of agro-ecological methods used	6.2	7.5**	5.1	<0.0001
Number of soil conservation methods used	4.5	4.9*	4.2	0.019
Pct of land with legumes	12	21**	4	0.0004
Fertilizers used amount, kg		109	192	n.a.
Amount of organic fertilizers purchased, kg	254	464**	15	0.0002
Manure use, kg per ha	15,821	21,190**	12,656	0.001
Number of crops	11.6	12.4*	11.0	0.033
Livestock diversity	4.7	5.2	4.2	0.128
% area with synthetic pesticides applied	n.a.	0	40	n.a.
% area with natural pesticides applied	n.a.	29	0	n.a.

Note: P-values denote a significant difference at 5% * and 1% ** levels.
Source: Qiao et al., 2009.

permitted organic inputs. This trend is demonstrated in case studies in China and Brazil (Oelofse et al., 2010b; cf. also Chapter 9). The farming systems, in general, did not undergo major changes in their cropping patterns, whilst there is a general heavy reliance upon input substitution for pest and soil fertility management (Oelofse et al., 2011).

According to Lotter (2003), conversion to OA typically requires structural changes to the farm rather than simple reduction or elimination of synthetic inputs from an otherwise conventional system. Rosset and Altieri (1997), in contrasting conventional agriculture with an agro-ecological approach, argue that the simple substitution of inputs may be able to ameliorate some environmental impacts of agriculture, yet does not reduce the vulnerability of monocultures, thus reducing the potential of sustainable agriculture. Guthman (2000) assessed to what degree a broad range of organic farmers in California practise the techniques of ecological agriculture and found that most farmers' practices fall short of agro-ecological ideals.

The IFAD review also found that, given time for transition, farmers could adapt to new practices and were generally able to secure sufficient organic nutrient inputs (from both on-farm and off-farm sources) and bio-pesticides from local sources (IFAD, 2005). It is therefore extremely important to recognize that conversion to organic agriculture is a long transition for farmers.

Using survey data collected from 973 organic farmers in three German regions during the spring of 2004 (Best, 2008), early and late adopters of organic farming are compared concerning farm structure, environmental concern, attitudes to organic farming, and membership in organic movement organizations. The results indicate that, on average, newer farms are more specialized and slightly larger than established ones and there is a growing proportion of farmers who do not share pro-environmental attitudes.

In the Wanzai, China, case (Qiao et al., 2009), the number of agro-ecological measures used was higher on farms with six and seven years of organic experience compared with farms that had been converted only two years before the interview. The same was found for the use of soil conservation methods. In Brazil, certified organic farmers required a long period to experiment and acquire knowledge of new management practices whilst their farming systems attain a new degree of resilience (Oelofse et al., 2011).

Oelofse et al. (2011) recognized that farmers' management practices are in transition and farmers operate within limits. They conclude their study by stating that the 'fulfilment of agro-ecological ideals, based upon system redesign to create bio-diversified cropping patterns and enhance pest management and nutrient cycling, is perhaps an end goal for organic farmers in a long transition'.

Market-oriented crop diversity and balance between building system resilience with economic considerations

The main driver of the rapid development of certified organic agriculture in developing countries is the export market, which exerts a strong influence upon agro–ecological changes occurring on organic farms. In particular, a group succinctly branded 'ecological entrepreneurs' has emerged (Parrott and Marsden 2002). Based upon new market opportunities, and often in collaboration with companies in the North, they seek out producers that are in a position to supply certain products according to specified standards (Parrott and Marsden, 2002). Organic companies typically help organize farmers, often within a contract farming model, and provide training and services.

Fraser (2006) recognizes the possible threat of the international trade focus resulting in ecologically fragile monocultures. Bakewell-Stone et al. (2008) analysed the potential of organic agriculture in Tanzania, and whilst recognizing a range of positive outcomes, they expressed concern about the strong international market focus upon certain high-value crops. Oelofse et al. (2011) demonstrate how organic companies influence certified organic farmers' agro-ecological practices in case studies in Brazil, China and Egypt. Although the adoption of organic agriculture induced a number of fundamental changes to farmers' management practices, the practices did not always align with organic principles. In particular, a 'niche' market crop focus influenced farm crop diversification and limited agro-ecological changes to the farm, thus failing to decrease farmers' reliance upon external inputs. Organic niche market crops with high-value influence organic farmers' management decisions, particularly regarding the prioritization of diversity in the cropping systems for agro-ecological purposes.

The mechanisms that have brought about the conversion of these farms to organic are primarily economic. Under such conditions, farms became organic as they could produce a certain high-value commodity organically. At the farm scale this process attracted farmers to convert to organic practices and perhaps this is why the presented farming practices do not fully live up to agro-ecological and organic principles. Thus, finding a balance between building long-term system resilience (e.g. through diversity and improved rotations and management of set-aside areas) with shorter-term economic considerations is a big challenge, particularly since farms in these case studies are typically certified organic due to market demand for one or two target crops. As also demonstrated by Schmutz et al. (2007), for a higher vegetable-cropping intensity (up to 90 per cent) a more sophisticated mix of short-term fertility-building and N-trapping crops will be needed and such rotations may require further external addition of green waste or livestock manure. In particular, farm management within contexts where there is little institutional support for transition requires a degree of pragmatism in order to find a balance between agro-ecological and economic requirements (Oelofse et al., 2010a; see also Box 4.1).

Case studies of farming systems in different regions

Box 4.1 Organic fruit production, Itapolis, Brazil, South America

The Itapolis case (Brazil) consists of organic fruit (primarily citrus) family-farmers who are members of a small cooperative. The cooperative, which is also Fairtrade certified, mainly exports concentrated orange juice, primarily to the EU, whilst they supply the domestic market with avocado, passion fruit and vegetables. About 25 per cent of the cooperative farmers are organic, whilst the remainder are conventional.

The fruit systems in Itapolis are perennial, thus changes in cropping patterns cover a longer temporal scale than one year. The cooperative promotes crop diversification based primarily upon the reduction of economic risk, although there are also clear motives for increasing system ecological resilience. Analysing crop diversity using a crop diversity index for Itapolis demonstrated this move towards crop diversification, as compared to the conventional sample which had a larger degree of orange crops in monoculture. At the time of conversion to organic farming, the majority of organic farms relied primarily on orange trees grown in monoculture. The cooperative that assists organic farmers has advised and assisted farmers in the diversification of their production to include other fruits such as mango and guava. The organic sample in Itapolis contained examples of farmers experimenting with agro-ecological design principles, such as green manuring and intercropping of legumes (e.g. Crotalaria and Mucuna) and other crops (e.g. maize and cassava) in between the rows of fruit trees when planting new plots. Brazilian Environmental Regulation (Law 4771/65) stipulates that for São Paulo 20 per cent of the farm should be set aside for ecosystem management. The mean proportion of land set aside on organic farms was 12 per cent (ranging from 0 to 30 per cent). The organic certification body in Itapolis requires that farmers pursue a reforestation of set-aside land, giving the farmers a limited period in which to achieve this (Oelofse, 2010).

Generally, farms have a few livestock, mostly for subsistence, and do not use manure from their own animals to fertilize the fruit plots. Nutrients are imported into the system in the form of composted manures, mainly from neighbouring farms and from the cooperative, which initiated a large composting project (using the residues of mango and guava after processing and mixing it with residues from trees pruned in urban areas). Following technical advice from the cooperative, organic farmers now grow various species of legumes between the fruit trees (primarily Crotalaria spp.) while the trees are younger than four years.

Box 4.2 Diversified organic crop production in Jiangxi, P.R. China, Asia

Jiaohu Township of Wanzai County sits in the western Jiangxi Province, with 847 hectares of farmland, among which 769 hectares are paddy fields and 78 hectares dry land. The main crops are rice, lily, ginger, sweet potato, soybean, peanut, garlic, radish, etc. The study was carried out in three representative organic villages with at least five years of organic experience. For comparison, three conventional villages were selected from a neighbouring town that have similar agro-ecological and socio-economic conditions to Jiaohu (Qiao et al., 2009).

The environmental advantages of organic agriculture in this case area are backed by the development and adoption of agro-ecological methods. This together with the ban on chemically produced pesticides as described in part above enhances the planned diversity and soil fertility and protects water resources against pollution. The combination of an absence of chemical pesticides, higher levels of weeds in the fields and use of such methods as mulching, composting and intercropping is known from the literature to benefit soil microbes, insects, bird life and probably reptiles. A large part of the land (40 per cent) in the conventional villages in Wanzai was treated with synthetic pesticides.

The strong involvement of the local public office of agriculture in Jiaohu led to a continuous test of alternative organic cash crops and adoption of agro-ecological methods such as intercropping and insect traps. Moreover, the conversion of whole villages with the certification controlled by the local authorities created possibilities for involving a large and increasing number of private food companies in the marketing of organic cash crops. The local public office acted as intermediary and had its own interest in organizing the production from the smallholder farmers in the villages in order to secure the procurement of the products according to contracts negotiated with the companies. Training and development was thus organized partly by the companies (for example a full-time extension worker was employed to promote strawberry production) and partly by the town office (also drawing on the agronomists at the county level).

Figures 4.2a and b Agro–ecological methods applied in organic production based in Jianxi, China

Box 4.3 Push–Pull system of maize pest management in Africa

A conservation agricultural approach known as 'Push-Pull' technology has been developed for integrated management of stemborers, striga weed and soil fertility. Push-Pull was developed by scientists at the International Centre of Insect Physiology and Ecology (icipe), in Kenya and Rothamsted Research, in the UK, in collaboration with other national partners. The technology is appropriate and economical for the resource-poor smallholder farmers in the region as it is based on locally available plants, not expensive external inputs, and fits well with traditional mixed-cropping systems in Africa. To date it has been adopted by over 46,000 smallholder farmers in East Africa where maize yields have increased from about 1 tons/ha to 3.5 tons/ha, achieved with minimal inputs. The technology involves intercropping maize with a repellent plant, such as desmodium, and planting an attractive trap plant, such as Napier grass, as a border crop around this intercrop. Gravid stemborer females are repelled or deterred away from the target crop (push) by stimuli that mask host apparency while they are simultaneously attracted (pull) to the trap crop, leaving the target crop protected. Desmodium produces root exudates some of which stimulate the germination of striga seeds and others inhibit their growth after germination. This combination provides a novel means of in situ reduction of the striga seed bank in the soil through efficient suicidal germination even in the presence of graminaceous host plants. Desmodium is a perennial cover crop (live mulch) which is able to exert its striga control effect even when the host crop is out of season, and together with Napier grass protects fragile soils from erosion. It also fixes nitrogen, conserves soil moisture, enhances arthropod abundance and diversity and improves soil organic matter, thereby enabling cereal cropping systems to be more resilient and adaptable to climate change while providing essential environmental services, and making farming systems more robust and sustainable.

Source: www.push-pull.net

For a further example, see Case study 3.

Indicators for use of agro-ecological practices assessment in organic farming

Why do we need indicators for agro-ecological practices assessment?

As described above, the principles of organic agriculture suggest and stipulate the use of various agro-ecological practices aimed at achieving goals for e.g. soil fertility, biological pest control, reduced dependence on fossil fuel and bio-diversity preservation. However, these values are only partially codified in rules and regulations, thereby allowing the compromising of a more holistic vision of organic farming (Milestad et al., 2008). The new European Regulation for organic production (EEC 834/2007 and EEC 889/2008) does include principles for organic production. However, not all are translated into production rules that can be part of inspection and certification (Padel et al., 2007). This mostly affects agro-ecological system values such as biodiversity and nutrient recycling, as well as the lack of social considerations (Padel et al., 2007; Lockie et al., 2006). Thus, the necessity of transparency in the interest of trade has made possible a rationalization and simplification of organic meanings (Tovey, 1997; Allen and Kovach, 2000). There are some examples of private labels introducing more specific rules for e.g. biodiversity preservation on organic farms, such as the Bio Suisse standard, which have minimum requirements for non-cultivated habitats (see below).

As shown previously, in reality not all organic farms are actually employing or developing agro-ecological practices and a focus on a few high-value cash crops or livestock products may lead to low levels of diversity and recycling and high dependence on external inputs. This development was observed in California and in Europe as reviewed by Darnhofer et al. (2010), who eventually concluded that 'despite the case studies reporting symptoms that have been linked to it, the available data is inadequate to confirm or to refute the con-ventionalization hypothesis in the European context. This is not least due to the fact that the variables used to identify the changes do not reliably indicate conventionalization'. Therefore, Darnhofer et al. (2010) suggested 'providing farmers with guidance on the development paths that are in line with the principles of organic farming'. To achieve this, an assessment framework whose indicators are based on the principles of organic farming would seem a useful tool.

Towards organic farming principle-based indicators

The goal of developing an assessment framework is for it to serve as a tool to assess and guide future developments of organic farming methods and practices. In this section we will introduce the idea of using sets of indicators for bench-marking among organic farmers the degree to which they follow and develop their practices in accordance with the principles. This could be applied to all aspects of organic principles including the fairness of labour use and business conditions, however in this chapter we only focus on indicators for organic farms' adherence to organic principles vis-à-vis agro-ecological aspects. In relation to IFOAM's set of principles this will especially be related to 'Health' and 'Ecology'.

Numerous indicator sets for assessing a farm's sustainability and environmental impact, animal welfare, etc. have been developed over the last 25 years, mainly reflecting the farms' contributions to external and societal expectations. We acknowledge that there is a large literature relevant to the proper development of indicators for agricultural sustainability with many good examples of coherence between objectives, criteria and indicators (Guthman, 2000; Rigby and Caceres, 2001; Castoldi and Bechini, 2010; Bockstaller et al., 1997). Moreover there are choices to be made in terms of indicator types (management-oriented vs. results-oriented; Halberg et al., 2005b), units (per ha, per kg), time frames and scales (field, farm, landscape levels), data recording and of the right tools to calculate indicator values and present them including benchmarking and other use of the information provided (Halberg et al., 2005a). We, therefore, do not propose a new, independent set of indicators. Rather, we suggest building on the existing indicator sets and tools to pick a subset specifically relevant for facilitating discussions with farmers and other stakeholders of how a specific farm or selected typical farming systems perform and may develop in relation to goals and principles for organic agriculture.

Darnhofer et al. (2010) suggest setting up indicators reflecting the organic principles and the degree to which certain practices on a farm are in contradiction with the principles. Examples are: 'Reliance on easily soluble (nitrogen) fertilizers'; 'Inadequate crop rotation or unbalanced crop sequence'; 'Prolonged and intensive use of plant protection products that are known to be problematic (e.g. sulphur, copper, pyrethrum)'; 'High share of feed that is purchased (industrially produced) rather than produced on the farm (or by neighbouring farms)'. The indicators are meant to identify where a farmer has reverted to finding 'conventional' solutions to challenges faced in production and management decisions. While the idea of using sets of indicators to guide the development of organic agriculture in accordance with the principles is sympathetic and necessary it may be more constructive to focus on indicators that demonstrate a positive compliance and development rather than 'the degree of conventionalization'. In this way, the development of such indicators can strengthen the integrity of the organic brand.

Table 4.2 lists the potential agro-ecological indicators related to the organic agriculture principles as adopted by IFOAM, with the main focus on 'Health' and 'Ecology'. The list is not intended to be final or comprehensive, and also excludes important aspects of fairness such as social and economic conditions for the farm family and workers. The indicators are selected according to agro-practices rather than their impacts. The latter would be more desirable, but the data are very rarely available. Information about farmers' agro-ecological practices are relatively easy to get during interviews with operators, although some of them are just rough and subjectively estimated such as the degree of on-farm nutrient management, although we hope they can be robust to reflect their impact on ecological processes and agro-biodiversity based on organic agriculture principles.

According to the discussion in the second section above and assessments by Guthman (2000) and Darnhofer et al. (2010), in order for crop production to comply with the principles of organic agriculture central agro-ecological measures are: diversifying crop patterns including crop rotation, intercropping and integration of husbandry; strengthening agro-biodiversity (crop diversity and share of natural habitats); on-farm nutrient management; on-farm biological pest and disease control; innovative weed management; and avoidance of the use of non-renewable resources and off-farm inputs. These six aspects can be regarded as criteria of organic farming; based on these criteria, a set of indicators are selected in Table 4.2.

Organic farmers' management practices are presented and analysed in six categories with different numbers of indicators: diversifying crop patterns (3), biodiversity enhancement (2), on-farm nutrient management (3), on-farm biological pest and disease management (3), innovative weed control practices (3), no restricted material (1); in total, there are 15 indicators. These indicators presented in Table 4.2 could be directly translated into standards to ensure practices are more agro-ecological. We can set up some reference value for implementation, like Bio Suisse has done.

Perspectives for adopting more agro-ecological measures in organic agriculture

Assessment of the practically applied agro-ecological measures

The evidence presented in this chapter suggests that agro-ecological practices could be implemented and promoted in organic agriculture; the adoption of organic agriculture has induced a number of fundamental changes to farmers' management practices. It is possible to organize organic cash crop production in global organic food chains in such a way that smallholder farmers are economically competitive and at the same time adopt agro-ecological methods, such as in Jiaohu, China (Box 4.2).

But for most of the organic farms, implementation of agro-ecological practices was limited compared with the principles. Research provides new knowledge of what types of changes in farming practices participation in global organic food chains might incur. Our findings show that organic farming does have a transformative potential like in Itapolis, Brazil (Box 4.1). Fulfilment of agro-ecological ideals, based upon system redesign to create bio-diversified cropping patterns and enhance pest management and nutrient recycling, is perhaps an end goal for organic farmers in a long transition.

The driver of this transformation is inextricably linked to the powers of the agro-industrial food chain. The farmers' crop choice is often heavily influenced by the organic market demand, thus causing farmers to focus more on target

Table 4.2 Agro-ecological indicators for degree of compliance with principles of organic agriculture

Criteria and related to IFOAM principles	Indicators	Rationale for selection
Diversifying crop patterns (principles of health and fairness)	Crop rotation of rational sequence with legume crops	Providing for a constant soil cover over time, nutrient balance with soil-enhancing crops and also good for pest and disease prevention
	Intercropping with share of legumes crops	Organic farming confirms all its potential and versatility in adapting to environmental and cultural context with different styles: mixed intercropping, intercropping in rows, strip intercropping, etc.
	Integration degree of plant production and husbandry	Recycling of wastes and by-products of plant and animal origin as input in plant and livestock production on farm level; regional level
Biodiversity enhancement (principles of health and ecology)	Crop diversity using Shannon Index	Wide range of crops and varieties should be grown to enhance the sustainability, self-reliance and biodiversity value of organic farms
	Share of natural habitats	Creating a heterogeneous environment outside of cultivated fields to pursue the stability of the ecosystem and its function such as productivity
On-farm nutrient management (principles of health, ecology and fairness)	Degree of on-farm fertility management	Different level green manure, legume crops and cover crops/mulching
	On-farm compost/manure	Treating the soil with farmyard manure/composting; crop residue return/straw residue return after feeding cattle
	Off-farm compost/manure or commercial input	Off-farm compost and permitted inputs are allowed provided the first two measures are not enough for fertility maintenance
On-farm biological pest and disease management (principles of health, and ecology)	Local and highly resistant variety	Using resistant crop genotypes.
	Degree of vegetation management and biological control	Hedgerows; trap crops/companion crops/natural enemy protection/physical traps/pheromone traps
	Application of approved pesticides	Pest and disease control using locally available materials such as neem, garlic, ashes, pheromone traps, pyrethrum and euphorbia.
Innovative weed control practices (principles of health and ecology)	Innovative weed control practices	Ways to reduce weed emergence (using crop sequencing and crop choice, cover crops and tillage) or reduce weed competition using cultural methods such as crop-type selection, planting pattern and fertilization strategy
	Manual weed control	Hand-weeding and hoeing are effective technologies, albeit extremely labour-intensive ones
	Mechanical control	Good way for weed control but energy consuming
No restricted material	Avoidance of restricted or controversial material	The materials or substances are potentially more toxic, environmentally problematic

crops than a holistic vision of their farm. Finding a balance between building system resilience through improved rotations and management of set-aside areas with economic considerations is a big challenge for organic farmers.

The form of organization to the production and marketing influences the degree of local adherence to organic principles and involvement in development of organic praxis building on agro–ecological principles. Moreover, the degree of facilitation by the contracting companies, NGOs or the public extension service influences the adoption of such methods and this is not always taken seriously by the companies involved.

Suggestions and perspectives

In order to be economically and ecologically resilient, energy and nutrient self-reliant as well as adaptive to climate changes and environmental issues and to optimize resource use, organic farming systems need to be diverse. However, organic agriculture is also under pressure to become specialized due to market demands and economic and farm management reasons. Therefore, the following suggestions and perspectives could be drawn based on the above evidence to improve the adoption of agro-ecological measures in organic agriculture. See also Chapter 10 for a social-movement approach to agro–ecology.

More feasible and principle-based rules for future regulations

All organic farms have to adhere to organic regulations and need to be certified with the minimum requirement, but not all aim for more than that, i.e. living up to the organic principles as closely as possible. Thus, it is possible for organic farms to focus on the regulations only, which in combination with other factors may in fact push a farm away from fulfilling the principles of organic farming.

The reformulation of the IFOAM organic principles is an open attempt to accommodate the challenges of the globalization of organic agriculture. An example of this type of regulation is the Bio Suisse standard 2009, which formulates measures to enhance biodiversity in organic farming. The standard states:

> It is the farm manager's duty to retain, complement or create near-natural habitats (ecological compensation areas), and to care for them in a professional manner. Ecological compensation areas must constitute at least 7 per cent of the agricultural area of the holding. The proportion of less intensively used meadows and extensive meadows, extensively used pasture, woodland grazings, or meadows mown for animal bedding must comprise at least 5 per cent of the permanent grassland. The buffer strips of hedges, copses, riparian woodlands may be included in the 5 per cent. A grassy margin of at least 0.5 m must be maintained alongside paths. In permanent crops, the first 3 m of such margins lying perpendicular to the main

direction of cultivation are counted as headlands and thus as part of the cultivated area.

By so integrating organic principles in rules or standards, farmers will implement more organic principles with the practical measures like Bio Suisse prescribes or under the São Paulo State Environmental Regulation (Law 4771/65) in the case study of Itapolis, Brazil (Box 4.1); it is feasible for farmers to carry out. The question is whether more principle- and result-based rules are a feasible, or desirable, option for future regulation (Oelofse, 2010).

Subsidies/policy support from government

Organic farming is both a technical management strategy and a political struggle. In the current situation it seems that if organic farming is to fulfil its aim of environmental safety, organic farming has to be pushed either by farmers and consumers or by governmental subsidies for desired practices and/or penalties for undesired practices. 'Being knowledge rather than input intensive, organics needs affirmative action from the state' (Morgan and Murdoch, 2000). Economic viability for organic farms can only be achieved through cooperation with market actors or governments to compensate farmers for the extra labour and lower yields. If this does not happen organic farming may soon experience conditions similar to industrial conventional agriculture where farmers' efforts and knowledge are replaced with fossil fuel, machines and external inputs.

Organic agriculture has also been promoted in developing countries as a strategy for rural development. The inclusion of small-scale farmers in global organic markets can provide a host of livelihood benefits. An increased focus on organic farmers' agro-ecological practices can be a way of reducing smallholder farmers' vulnerability by building more resilient and robust farming systems.

Empowerment of organic farmers through influence from organic cooperatives

Organic farming in developing countries typically involves small householders. Organic companies/cooperatives will influence certified organic farmers' agro-ecological practices as shown in the case studies in Brazil and China (Boxes 4.1 and 4.2; Oelofse et al., 2010a).

Organic farming is primarily knowledge-intensive, whereas conventional farming is more chemical- and capital-intensive; organics can therefore be an advantage for poorer farmers because it reduces capital needs and debt risks (Panneerselvam et al., 2011). However, organic agriculture requires training and capacity building and certain inputs (seeds, compost, etc.) all of which are limiting factors if they are not supplied by organizations or companies. Accordingly, it is difficult to establish a one-size-fits-all approach, since conditions will vary in different zones. Organic projects require that time be built into the process for farmers to test and learn new technology and methods. Knowledgeable extension services are critical. Local know-how, especially from experienced farmers and knowledgeable elders, can smooth the transition and reduce risks. It is also important to provide farmers with good access to sources of knowledge about the application of organic methods to their crops and agro-ecological conditions.

Smallholder farmers should be empowered on the benefit of agro-ecology for sustainability of their production systems; by improved training and extension services based on both local and introduced knowledge tailored to smallholder farms. Providing farmers with technical assistance to build capacity for organizational and legal skills is required in order to foster more sustainable, as well as agro-ecologically sound, production systems.

Research on new mixed-farming systems with multiple objectives

Research should develop and document innovative collaboration between specialized farmers to achieve new mixed-farming systems with multiple objectives. Crop and livestock systems should be redesigned to increase integration, improve the cycling of nutrients and organic matter, and diversify the production systems using strategies such as mixed cropping, agro-forestry, integration of grassland and rangelands into cash crop systems. Research will also include optimization of cycles at a regional scale such as high-quality household waste, human urine, bioenergy, ashes, etc. Transition patterns should also be considered in order to define possible paths for farming systems' evolution towards more diversification. Research may include the perspective of whole-food systems using value-chain and life cycle assessment methodologies and address aspects of food quality.

Conclusions

Organic agriculture is intended to put the principles of environmentally friendly development into effect through the realization of agro-ecosystems that provide for maximum utilization of local resources within the production context. Agro-ecological practices of organic farms are to be considered optimal for sustainability when all the principles are put into practices. In reality, not all the principles can be realized especially under the influence of organic market demand in the agro-industrial food chain; the balance between building system resilience and economic considerations is a big challenge for organic farmers.

In this chapter a set of potential agro-ecological indicators related to the organic agriculture principles is put forward to assess the organic farmers' management practices which can guide the development of organic agriculture in accordance with the principles and demonstrate a positive compliance and development.

In order to promote the implementation of agro-ecological measures in organic farming, innovative ways should be considered and measures should be carried out such as prescribing more principle-based and feasible rules for future regulations; providing more subsidies/policy support from government or NGOs to guarantee the adoption of agro-ecological practices; empowering organic farmers through influence from organic cooperatives and developing new mixed-farming systems with multiple objectives.

Note

1   The studies behind parts of Chapter 4 and Case Study 4 were supported by the Asian Development Bank Institute, Tokyo and the Danish Ministry for Food and Agriculture. The opinions expressed in the paper are the responsibility of the authors alone. We thank Dr Sununtar, ADB for valuable input and inspiration.

References

Allen, P. and Kovach, M. (2000) 'The capitalist composition of organic: the potential of markets in fulfilling the promise of organic agriculture', *Agriculture and Human Values*, 17, pp. 221–232

Altieri, M. (1995) *Agroecology: The Science of Sustainable Agriculture*, 2nd edn, Westview Press, Boulder, CO

Altieri, M. (2002) 'Non-certified organic agriculture in developing countries', in N.E.H. Scialabba and C. Hattam (eds) *Environmental and Natural Resources Series*, FAO, Rome, pp. 107–138

Badgley, C., Moghtader, J.K., Quintero, E., Zakem, E., Chappell, M.J., Avilés, K.R. et al. (2007) 'Organic agriculture and the global food supply', *Renewable Agriculture and Food Systems*, 22, 2, pp. 86–108

Bakewell-Stone, P., Lieblein, G. and Francis, C. (2008) 'Potentials for organic agriculture to sustain livelihoods in Tanzania', *International Journal of Agricultural Sustainability*, 6, pp. 22–36

Best, H. (2008) 'Organic agriculture and the conventionalization hypothesis: a case study from West Germany', *Agriculture and Human Values*, 25, pp. 95–106

Bio Suisse (2009) *Bio Suisse Standards for the Production and Marketing of Produce from Organic Farming*, Bio Suisse, Basel, available at: http://organicstandard.com.ua/files/standards/en/suisse/rl_2009_e.pdf

Bockstaller, C., Girardin, P. and van der Werf, H.M.G. (1997) 'Use of agroecological indicators for the evaluation of farming systems', *European Journal of Agronomy*, 7, pp. 261–270

Caporali, F. (2004) *Agriculture and Health: The Challenge of Organic Agriculture.* EDITEAM sas, Cento (FE), Italy

Castoldi, N. and Bechini, L. (2010) 'Integrated sustainability assessment of cropping systems with agroecological and economic indicators in northern Italy', *European Journal of Agronomy*, 32, 1, pp. 59–72

Darnhofer, I., Lindenthal, T., Bartel-Kratochvil, R. and Zollitsch, W. (2010) 'Conventionalisation of organic farming practices: from structural criteria towards an assessment based on organic principles, A review', *Agronomy for Sustainable Development*, 30, pp. 67–81

Eyhorn, F. (2007) 'Organic farming for sustainable livelihoods in developing countries: the case of cotton in India', PhD diss. Department of Philosophy and Science, University of Berne http://www.zb.unibe.ch/download/eldiss/06eyhorn_f.pdf (accessed February 12, 2009)

Fraser, E.D.G. (2006) 'Crop diversification and trade liberalization: linking global trade and local management through a regional case study', *Agriculture and Human Values*, 23, pp. 271–281

Guthman, J. (2000) 'Raising organic: an agro-ecological assessment of grower practices in California', *Agriculture and Human Values*, 17, pp. 257–266

Halberg, N., Verschuur, G. and Goodlass, G. (2005a) 'Farm level environmental indicators, are they useful? A review of green accounting systems for European farms', *Agriculture, Ecosystems and Environment* 105, pp. 195–212

Halberg, N., van der Werf, H.M.G., Basset-Mens, C., Dalgaard, R. and de Boer, I.J.M. (2005b) 'Environmental assessment tools for the evaluation and improvement of European livestock production systems', *Livestock Production Science*, 96, pp. 33–50

IFAD (2002) 'Thematic evaluation of organic agriculture in Latin America and the Caribbean'. Rep. EC 2002/32/W.P.3., International Fund for Agricultural Development (IFAD), Rome

IFAD (2005) 'Organic agriculture and poverty reduction in Asia: China and India focus'. Thematic Evaluation. Rep. 1664., International Fund for Agricultural Development (IFAD), Rome

IFOAM (2005) *The IFOAM Basic Standards for Organic Production and Processing*, IFOAM, available at: http://www.ifoam.org/about_ifoam/standards/norms/norm_documents_library/IBS_V3_20070812.pdf

Lockie, S., Lyons, K., Lawrence, G. and Halpin, D. (2006) *Going Organic. Mobilizing Networks for Environmentally Responsible Food Production*, CABI Publishing, Wallingford

Lotter, D.W. (2003) 'Organic agriculture', *Journal of Sustainable Agriculture*, 21, pp. 59–128

Love, B. and Spaner, D. (2007) 'Agrobiodiversity: its value, measurement, and conservation in the context of sustainable agriculture', *Journal of Sustainable Agriculture*, 31, pp. 53–82

Lyngbæk, A.E., Muschler, R.G. and Sinclair, F.L. (2001) 'Productivity and profitability of multistrata organic versus conventional coffee farms in Costa Rica', *Agroforestry Systems*, 53, pp. 205–213

Mäder, P., Fliebach, A., Dubois, D., Gunst, L., Fried, P. and Niggli, U. (2002) 'Soil fertility and biodiversity', *Organic Farming Science*, 296, n. 5573

Milestad, R., Wivstad, M., Lund, V. and Geber, U. (2008) 'Goals and standards in Swedish organic farming: trading off between desirables', *International Journal of Agricultural Resources, Governance and Ecology*, 7, pp. 23–39

Morgan, K. and Murdoch, J. (2000) 'Organic vs. conventional agriculture: knowledge, power and innovation in the food chain', *Geoforum*, 31, pp. 159–173

Oelofse, M. (2010) 'The sustainablility of organic farming in a global food chains perspectives – the agroecology of organic farming system', PhD thesis, SL Grafik

Oelofse, M., Høgh-Jensen, H., Abreu, L.S., Almeida, G.F., Hui, Q.Y., Sultan, T. and de Neergaard, A. (2010a) 'Certified organic agriculture in China and Brazil: market accessibility and outcomes following adoption', *Ecological Economics*, 69, 9, pp. 1785–1793

Oelofse, M., Høgh-Jensen, H., Abreu, L.S., Almeida, G., El-Araby, A., Hui, Q. and de Neergaard, A. (2010b) 'A comparative study of farm nutrient budgets and nutrient flows of certified organic and non-organic farms in China, Brazil and Egypt', *Nutrient Cycling in Agroecosystems*, 87, pp. 455–470

Oelofse, M., Høgh-Jensen, H., Abreu, L.S., Almeida, G.F., El-Araby, A., Hui, Q.Y., et al. (2011) 'Organic farm conventionalisation and farmer practices in China, Brazil and Egypt', *Agronomy for Sustainable Development*, doi 10.1007/s13593-011-0043-z

Padel, S., Roecklingsberg, H., Verhoog, H., Alroe, H., De Wit, J., Kjeldsen, C. and Schmid, O. (2007) 'Balancing and integrating basic values in the development of organic regulations and standards: proposal for a procedure using case studies of conflicting areas', University of Wales, Aberystwyth, available online at: http://orgprints.org/10940/

Panneerselvam, P., Hermansen, J.E. and Halberg, N. (2011), 'Food security of small holding farmers: comparing organic and conventional systems in India', *Journal of Sustainable Agriculture*, 35, 1, pp. 48–68

Parrott, N. and Marsden, T. (2002) 'The real Green Revolution: organic and agroecological farming in the South', Greenpeace Environmental Trust, UK

Parrott, N., Olesen, J.E. and Høgh-Jensen, H. (2006) 'Certified and non-certified organic farming in the developing world', in Halberg, N., et al. (eds), *Global Development of Organic Agriculture: Challenges and Promises.* CABI Publishing, Wallingford, pp. 153–179

Perfecto, I. and Vandermeer, J.H. (2008), 'Biodiversity conservation in tropical agroecosystems: a new conservation paradigm', *Annals of the New York Academy of Sciences*, 1134, 1, pp. 173–200

Pimentel, D., Hepperly, P., Hanson, J., Seidel, R. and Douds, D. (2005) 'Environmental, energetic, and economic comparisons of organic and conventional farming systems', *Bioscience*, 55, 7, pp. 573–582

Qiao, Y., Setboonsarng, S. and Halberg, N. (2009) 'PRC country study on organic agriculture and the Millennium Development Goals', Asian Development Bank Institute Working Paper.

Rigby, D. and Caceres, D. (2001) 'Organic farming and the sustainability of agricultural systems', *Agricultural Systems*, 68, pp. 21–40

Rosset, P.M. and Altieri, M.A. (1997) 'Agroecology versus input substitution: a fundamental contradiction of sustainable agriculture', *Society & Natural Resources*, 10, pp. 283–295

Schmutz, U., Rayns, F. and Firth, C. (2007) 'Balancing fertility management and economics in organic field vegetable rotations', *Journal of the Science of Food and Agriculture,* 87, pp. 2791–2793

Scialabba, N.E.-H. and Hattam, C. (2002) *Organic Agriculture, Environment and Food Security.* Food and Agriculture Organization of the United Nations (FAO), Rome. www.fao.org/docrep/005/y4137e/y4137e00.htm

Stockdale, E.A., Lampkin, N.H., Hovi, M., Keatinge, R., Lennartsson, E.K.M., Macdonald, D.W. et al. (2001) 'Agronomic and environmental implications of organic farming systems', *Advances in Agronomy*, 70, pp. 261–262

Tovey, H. (1997). 'Food, environmentalism and rural sociology: on the organic farming movement in Ireland', *Sociologia Ruralis*, 37, 1, pp. 21–37

5 The potential of organic agriculture for contributing to climate change adaptation

Adrian Muller, Balgis Osman-Elasha and Lise Andreasen

Introduction

Agriculture is and will be affected negatively by climate change in many regions. Most vulnerable are poor agricultural communities in the Global South and assuring food security for the twenty-first century requires successful adaptation to climate change in agriculture. Besides being affected by climate change, agriculture also contributes significantly to it. Some 10–15 per cent of total global greenhouse gas emissions are direct agricultural emissions. Adding emissions from conversion to agricultural land increases this share to more than 30 per cent.

This chapter addresses the potential of organic farming systems to adapt to climate change. Depending on the context, the focus lies on the adaptation of the agricultural production system, of a community or at farm level. Adaptation strategies can be as diverse as changes in varieties cultivated (e.g. with increased drought resistance) or in management practices (e.g. mulching instead of burning of crop residues to reduce irrigation needs), physical investments for improved water management, a change in livelihoods from crop farming to animal husbandry, or even migration. Assessing success of adaptation can take several years to decades and some initially successful adaptation strategies may turn out to be a failure later on. Furthermore, the concepts used to frame adaptation in agriculture, such as resilience, need some further clarification and discussion.

In this chapter, first we review briefly the main impacts of climate change on agriculture. We then present and clarify the key concepts for discussing adaptation measures in agriculture and address key challenges for their implementation. Subsequently, we analyse the adaptation potential of organic agriculture. We find that organic agriculture addresses most of the important challenges of successful adaptation. We also find that the current adaptation discussion in general does not sufficiently include the full adaptation potential of soil fertility (namely high levels of soil organic matter), of crop rotations and of biodiversity. After this, mitigation in organic agriculture is addressed briefly, in close linkage to the adaptation discussion. This is focused on the mitigation aspects that provide livelihood strategies and on synergies with adaptation, as there is already an extensive literature on the physical mitigation potential of organic agriculture. Mitigation aspects are also illustrated in Case study 3.

Predicted impacts of climate change on agriculture

The impact of climate change on agriculture can be summarized as follows. For further detail see Easterling et al. (2007), Meehl et al. (2007) and Rosenzweig and Tubiello (2007).

- CO_2 levels, temperatures and climate variability and the frequency of extreme events will increase.
- Precipitation, transpiration regimes, crop growing seasons and weed, pest and pathogen pressure will change, but the direction of these changes can vary per region.
- Impacts on global agricultural production will be minor in the first half of the century and increasingly negative afterwards.
- Negative impacts will however be stronger and earlier in low-latitude developing countries than in mid- and high-latitude developed countries.

In principle, higher CO_2 levels generally lead to higher production. The magnitude of this effect depends on the plant species, though. These positive effects are, however, counterbalanced by adverse effects of increasing temperatures. Moderate increases may also increase production, but higher increases beyond 1.5°C decrease production. Increased climate variability and extreme events have also largely negative effects on agricultural production. Heavy rain and floods, droughts, storms, fires and pest outbreaks are examples of such extreme events. Such extreme events can lead to additional damages beyond the adverse impacts to be expected from changes in mean temperature and precipitation alone.

Precipitation and transpiration regimes will change, and general patterns are shifting from the low-latitude dry zones to higher latitudes, and an increase in precipitation in high latitudes and around the equator. Globally, precipitation will increase, but regional variation is considerable. Climate change will also change the current monsoon and El Niño patterns. Changing precipitation patterns may have large effects on water availability and irrigation water requirements. Changes in weed, pest and disease patterns manifest in a spread from low to mid-latitudes with warming. In addition, climate extremes will promote plant disease and pest outbreak. Plant damage from pests will be strongly determined by the effects of the interactions between increases in CO_2 levels, temperature and precipitation.

Specific impact on livestock stems from additional losses due to droughts and extreme events. Increased temperatures and CO_2 levels will change the species composition of pastures and also nutrient contents of forage. Whether this has positive or negative impacts is not yet clear, though.

A specific aspect regarding soil carbon sequestration is the fact that increased temperature and precipitation can interact to change soils from carbon sinks to carbon sources. Several climate models predict such behaviour around 2050, and its occurrence would have important positive feedback effects on global climate change. In addition, soils changing from sinks to sources would challenge the mitigation potential in agriculture and the other benefits from soil carbon sequestration. These effects are very uncertain, though (Meehl et al., 2007).

Adaptation in an agricultural perspective

From predictions of future climate change impacts (see the preceding section) we derive several key messages for adaptation in agriculture. First, *impacts vary strongly per region*. This has to be accounted for when discussing adaptation. Some regions are affected positively, some negatively. In any case, production conditions are likely to change, necessitating some adaptation for optimal production even in favourable cases. Second, *water will be a key issue*, in particular due to water scarcity and drought, but also because of extreme precipitation events, waterlogging and flooding. Third, *increased weed, pest and disease pressure* will challenge agriculture. Fourth, *extreme events* will put further stress on agricultural production. Fifth, climate *variability* and the *risk* in agricultural production will increase. Finally, impacts affect people well *beyond agricultural production*, as this provides livelihoods for hundreds of millions of people further down the value chains as well.

Definitions and concepts related to adaptation

We discuss the important key concepts of adaptation, framing them in an agricultural context and synthesizing them within the rural livelihood framework (see Chapter 1 for a detailed discussion of the livelihood framework). This provides some clarification of the concepts, which usually is lacking in the discussion of adaptation in agriculture. This then serves as a basis to assess the adaptation potential of organic agriculture further down. The literature offers several definitions for these key concepts and related concepts and discussing various aspects of these. In the relevant core aspects, these definitions are often very similar. We decided to use the IPCC definitions from the Fourth Assessment Report 2007, wherever possible and expedient.

a) In the adaptation discussion, the *entities* of concern are often communities or socio-economic systems (but also households, individuals, etc.). We use the following definitions: '*Community* [. . . means] some definable aggregation of households, interconnected in some way, and with a limited spatial extent' (Smit and Wandel, 2006), and '[a] "*social-ecological system*" . . . encapsulates ecosystems and their human use by communities and institutions' (Osbahr et al., 2010).

b) The IPCC defines *adaptation* as '[a]djustment in natural or human systems in response to actual or expected climatic stimuli or their effects, which moderates harm or exploits beneficial opportunities' (IPCC, 2007: 869).

Smit and Wandel (2006: 282 ff.) emphasize that there are many forms of adaptations, which can be classified in many ways. Examples of categories for classification are the 'timing relative to stimulus (anticipatory, concurrent, reactive), intent (autonomous, planned), spatial scope (local, widespread) and form (technological, behavioural, financial, institutional, informational)'. They also emphasize the importance of the 'degree of adjustment or change required from (or to) the original system (Risbey et al., 1999). For an agricultural system facing water shortage exposures, a simple adaptation might be to use more drought resistant cultivars. A more substantial adaptation might be to shift away from crop farming to pastoralism. An even more substantial adaptation might be to abandon farming altogether.' See also the discussion of 'transformation' in point e) below.

Nelson et al. (2007) point out that the adaptation discussion is overly actor, actions and agency centred and that systemic views are neglected. The actor-centred approach is often based on an assessment of specific risks and the corresponding assessment of adaptive actions remains static, implicitly building on the assumption that systems largely are and remain in equilibrium. The systemic approach acknowledges that many systems are in a state of ongoing change rather than in equilibrium. Systems should be managed for flexibility and not for maintaining stability. In recent years, this focus on actors has been changing in some contexts. The framework document on adaptation from the FAO (2011), for example, addresses many systemic and long-term aspects (e.g. by its focus on food security and an ecosystem approach).

While the actor-centred approach focuses on the processes of decisions, the system-centred approach focuses on the effects of these processes on the system. The temporal component is crucial for the assessment of adaptation in the system-centred approach. Adaptation is not seen primarily in terms of a set of activities, but rather in how the activities 'feedback, either positively or negatively, into the system as a whole through time'. As adaptation should focus on maintaining flexibility, it has to be evaluated how current adaptation affects future flexibility. Different to the actor-centred adaptation approaches that 'foster adaptation that will lead to a state in which the social-ecological system deals effectively with perceived risks', the system-centred approach 'promotes managing the capacity of the system to cope with future change' (Nelson et al., 2007: 407).

Nelson et al. (2007: 407) emphasize that a systemic approach also requires a combined view on social and ecological systems as coupled systems. 'In this sense, a society may be able to cope well with change from a social perspective (e.g. improving irrigation technology and increasing agricultural subsidies), but an evaluation of overall resilience must also include the sustainability of the adaptation from an ecological perspective (e.g. the ecological impacts of increased farming and groundwater pumping).'

c)   *Adaptive capacity* is defined as 'The ability of a system to adjust to climate change (including climate variability and extremes) to moderate potential damages, to take advantage of opportunities, or to cope with the consequences' (IPCC, 2007). Similar definitions and a host of similar concepts (e.g. adaptability, coping ability, management capacity, stability, robustness, flexibility, resilience) are used in the literature (e.g. Smit and Wandel, 2006: 287; Nelson et al., 2007: 407).

Adaptive capacity increases with endowments of the various types of livelihood capital and with the quality of structures and processes, it is very context-specific and depends on characteristics at various scales (Smit and Wandel, 2006). Smit and Wandel (2006), for example, state that ' the capacity of a household to cope with climate risks depends to some degree on the enabling environment of the community, and the adaptive capacity of the community is reflective of the resources and processes of the region'.

Nelson et al. (2007) identify three fundamental characteristics for adaptive capacity, namely (1) sensitivity to change while still retaining structure and function (a point that has to be challenged, see the next issue 'd) resilience'); (2) self-organization; and (3) the capacity for learning. Based on an assessment of several studies on the determinants of adaptive capacity, they also conclude that endowment with various types of livelihood capitals, good governance structures and efficient institutional processes are key.

d) *Resilience* is '[t]he ability of a social or ecological system to absorb disturbances while retaining the same basic structure and ways of functioning, the capacity for self-organisation, and the capacity to adapt to stress and change' (IPCC, 2007: 880). As for example Nelson et al. (2007) point out, '[m]any definitions conflate resilience and adaptive capacity in part or in whole'. They rightly state that this is dangerous, as there can be highly resilient systems that are in undesirable states. In desirable states, resilience can support adaptive capacity, but as soon as a resilient state becomes undesirable, resilience is a hindrance to adaptation, as transformation would be needed.

e) *Transformation*, as opposed to gradual changes, is 'a fundamental alteration of the nature of a system once the current ecological, social, or economic conditions become untenable or are undesirable' (Nelson et al., 2007: 399). 'Transformability' is then '[t]he capacity to create a fundamentally new system when ecological, economic, or social (including political) conditions make the existing system untenable' (Walker et al., 2004).

f) Finally, we present 'exposure', 'sensitivity' and 'vulnerability' as three concepts determining how changes impact adapting entities. *Exposure* refers to the likelihood that a system experiences certain conditions (e.g. a drought) (e.g. Smit and Wandel, 2006). '*Sensitivity* is the degree to which a system is affected, either adversely or beneficially, by climate variability or change. The effect may be direct (e.g. a change in crop yield in response to a change in the mean, range or variability of temperature) or indirect (e.g. damages caused by an increase in the frequency of coastal flooding due to sea-level rise)' (IPCC, 2007: 881). '*Vulnerability* is the degree to which a system is susceptible to, and unable to cope with, adverse effects of climate change, including climate variability and extremes. Vulnerability is a function of the character, magnitude, and rate of climate change and variation to which a system is exposed, its sensitivity and its adaptive capacity' (IPCC, 2007).

These concepts are combined as follows. Climate change necessitates *adaptation* in agricultural *entities* such as communities or systems (but also households, families, individuals, etc.). How much adaptation is necessary depends on the *exposure, sensitivity* and *vulnerability* of the *adapting entity*. How well it can adapt is captured by its *adaptive capacity*. As long as changes and adaptation needs are gradual, *resilience* is an important aspect of adaptive capacity. Beyond certain levels of change and adaptation needs, transformation may be the alternative pathway and *transformability* the necessary characteristic of a system to pursue this. Adaptation measures can thus be judged based on their performance regarding exposure, sensitivity, vulnerability, resilience and transformation in the face of the main impacts listed above (see Table 5.1).

Linking this to the livelihood (LH) approach as presented in Chapter 1 makes it clear that adaptation is just the LH actions taken in the face of risks, opportunities and changes in the context related to climate change. Many authors on adaptation point out the importance of building up the various types of livelihood capital and establishing good institutional frameworks and good governance (e.g. Walker et al., 2004, 2006; Olsson et al., 2004; Nelson et al., 2007). The specific challenges to adaptation discussed below, such as scale, equity, barriers, learning, etc., also directly link to key aspects in the LH framework. In the livelihood approach, the adapting entity is primarily the household, but the concepts presented above point to the importance of adaptation in the context as well. Similarly, the differentiation of exposure, sensitivity and vulnerability adds further detail to the risk and opportunities used in the LH framework. This exemplifies how the LH approach is a broadly applicable basic structure that can be amended by the specific characteristics of specific challenges. But the amended LH approach as described in Chapter 1 also inspires and refines the adaptation discussion, as it points to the importance of power relations and the 'inner realities', such as the individual and family orientation, the inner human space and the emotional base, for example, as drivers behind actions.

Particular challenges of adaptation in agriculture

Although adaptation is common to all social systems and continuously taking place, it gained attention in the climate change debate only recently. Thus, adaptation is not new, but in the context of climate change, the mere scale and number of systems that need to adapt, the expected fast rate of change and the availability of a huge amount of information make adaptation to climate change a special type of adaptation (Fuessel, 2007). Adaptation to climate change is highly diverse and it is hardly ever done with respect to climate change alone. It is always taking place in the context of other developments in a system. Still, adaptation projects in agriculture are in many aspects not different from other resource management, rural development or institution-building projects. The following aspects pose particular challenges to adaptation to climate change.

a) *Scale* and the interactions between different scales matter greatly. As Burton and Lim (2005) and Olsson et al. (2004) point out, large-enough spatial and temporal scales need to be assessed for successful adaptation processes (e.g. national, not only local, respondent eco-system level), and criteria for success of adaptation also need to be chosen differently at different scales (Adger et al., 2005). Nelson et al. (2007) point out that adaptation can have adverse effects on parts of a community while it is beneficial to others. Similarly, adaptation can have adverse effects in the long run although it may appear successful in the short run. They emphasize that adaptive capacity depends on the temporal scale and frequency of perturbations, the spatial scale of perturbations, and the scale of organizational processes in relation to these perturbations. It is important that the scales of change or perturbations and of the corresponding adaptation processes match (Cumming et al., 2006). Replacing local within-community adaptive capacity with support from outside the boundaries of this community can pose a problem, as it can increase vulnerability and reduce adaptive capacity.

b) Albeit often being judged favourably in the literature and being used as defining for adaptive capacity or success in adaptation (e.g. Osbahr et al., 2010; Folke et al., 2002), (increased) *resilience* is not necessarily advantageous. It is a neutral concept and resilience can also be a hindrance to adaptation (if a system is captured in an undesirable state and transformation is needed, for example, see e.g. Walker et al., 2006). We thus disagree with Folke et al. (2002: 438) who state that '[w]hen massive transformation is inevitable, resilient systems contain the components needed for renewal and reorganization.'

The trade-off between resilience and adaptive capacity may become effective in several ways. Increased resilience at one location or scale may lead to increased vulnerability at another (e.g. minimum prices for some product, e.g. wool, that increase the resilience of individual farmers, but can lead to overproduction, accumulation and oversupply with detrimental effects on the economy and the individual farmers at a later point of time; Walker et al., 2006). Increased resilience towards certain specific or regular shocks can make a system more vulnerable to other, unexpected and new shocks (Walker et al., 2006; Osbahr et al., 2010). Adaptive capacity can also suffer from increased efficiency of a system in the sense of reduced redundancy. This can lead to reduced diversity in possible responses to perturbations (Walker et al., 2006).

c) When transformations occur, the *adapting entity* (e.g. the community) may disappear, as it has to be changed during transformation (e.g. from a community to a household focus in the context of new communities, if migration becomes an issue). This is usually not mentioned in the literature but it is a crucial aspect of transformability. This possibility makes the assessment of the entity that adapts much more difficult, and not allowing for changing it or for letting it disappear may foreclose promising adaptation strategies.

d) The presence of different effects on different parts of a system (see earlier discussion of 'scale') touch on issues of *equity*. One aspect of this is the fact that adaptation can reduce the vulnerability of those able to take advantage of governmental institutions, while already more vulnerable and marginalized parts of the system may stay vulnerable or even incur increasing vulnerability. Equity principles should thus be used and power relations and imbalances should be identified when assessing vulnerability and adaptation of a community (Nelson et al. 2007). Therefore it should always be asked 'who decides what should [adapt or] be made resilient to what, for whom [adaptive capacity or] resilience is managed, and to what purpose?' (adapted from Lebel et al., 2006 (ref. 106) in Nelson et al., 2007). Similarly, decisions often have to be taken regarding which level of adaptation has to be supported, and which levels of vulnerability are acceptable for which parts of a system. The discussion and negotiation of these issues requires management abilities and good governance and institutions.

e) Potential *barriers* to adaptation need particular attention. Adger et al. (2007) observe that high adaptive capacity does not necessarily lead to successful adaptation. Moser and Ekstrom (2010) address this and present a categorization and also concrete suggestions on how to best identify barriers. They also identify four key aspects that are of importance throughout many adaptation processes and bear the potential of barriers: leadership issues; availability of resources of any kind (physical, but also financial, informational, etc., – thus largely coincident with availability of livelihood capitals); communication and information; and values and beliefs.

f) *Learning* is crucial. Tschakert and Dietrich (2010) still observe a lack of learning tools and poor understanding of the specific aspects of learning and communication in adaptation. Access to learning tools, information, knowledge networks and related institutions remain poor, particularly in vulnerable regions and communities. The adaptation discourse does still not fully acknowledge the potential of adaptive and anticipatory learning but remains 'predominantly focused on responding to the predicted impacts of future climate change rather than addressing the underlying factors that determine chronic poverty, vulnerability, and adaptive capacity' (Tschakert and Dietrich, 2010).

g) *Planning, implementation* and *evaluation* of adaptation poses considerable challenges. Given the timescale of climate change, assessing success of adaptation has to be done over decades. This long-term aspect is a challenge to planning, implementation and evaluation, as they need to be flexible enough to adequately react to unforeseen changes. The timescale also tightly interlinks these aspects, as the evaluation for a certain period affects planning and implementation for the next and all future ones.

Due to the complexity of adaptation, planning, implementation and evaluation always depend on some approach on how to reduce complexity and how to structure the various aspects. Among others, Nelson et al. (2007), Adger et al. (2005) and Walker et al. (2006) present some possibilities for this. Fuessel (2007) suggests differentiating between hazard-based and vulnerability-based approaches. The former is most adequate to identify the current situation and raise awareness of the problem, while the latter serves to identify priority areas of action. Osbahr et al. (2010) concretely ask 'What is success [in adaptation]?' and point out the importance and difficulty of determining desirable vs. undesirable states of a system. Employing a resilience-based framework, they define 'the process of successful adaptation as that which increases system resilience but also, giving explicit treatment to governance, as that which promotes legitimate institutions to generate and sustain collective action'. We take a broader view than on resilience only (see above), but this quote indicates the crucial role of a systemic view for planning, implementation and evaluation. Nelson et al. (2007) refer to the literature (their ref. 111: Carlsson and Berkes, 2005), emphasize a dynamic element and thus frame this as follows: 'evaluation of adaptation management should focus on process rather than results and on function rather than structure'. This is also captured in the notion of 'adaptive comanagement' as discussed in detail in Olsson et al. (2004). Or in the metaphor of the 'adaptive cycle' that is adequate for many systems (growth, conservation, disturbance, reorganization), as described in Walker et al. (2006).

h) Finally, we thus want to explicitly emphasize the *systemic aspect* of adaptation again. Adaptation is not viable, when the ecological and the social sphere remain separated and when it is a mere combination of single practices and discrete and unrelated policy and technology decisions (Nelson et al., 2007; Folke et al., 2002; Tschakert and Dietrich, 2010). Osbahr et al. (2010) draw similar conclusions from their case-study analysis, where social network aspects and informal institutions play a key role in the success of adaptation. However, this does not mean that simplifications are impossible. The complexity of socio-economic systems can often be reduced considerably by identifying a few key variables that govern the main behavioural patterns of the system (Walker et al., 2006: proposition 4). Identifying those and understanding their linkages is far from simple, but taking this aim as a starting point can greatly facilitate analysis.

Adaptation in organic agriculture

In this section, we discuss the potential of organic agriculture for successful adaptation. We first compare climate change impacts in agriculture to specific characteristics of organic agriculture that support adaptation to them. We then take up the challenges of adaptation and assess how organic agriculture performs regarding those impacts, identifying promising aspects and areas for improvement.

The adaptation potential of organic agriculture

Organic agriculture offers promising answers to most of the main impacts of climate change on agriculture and their characteristics as identified above, namely that (1) impacts vary strongly per region; (2) water is a key issue (scarcity, extreme precipitation, etc.); (3) disease and pest pressure increases; (4) the frequency of extreme events (drought, precipitation, etc.) increases; (5) climate variability and the risk in agricultural production increase; (6) people outside of the agricultural sector are affected. Earlier papers address the adaptation potential of organic agriculture (Milestad and Darnhofer, 2003; Borron, 2006; Niggli, 2009; Scialabba and Müller-Lindenlauf, 2010), but they do not locate this in an encompassing adaptation framework. The latter three link climate change impacts directly to possible adaptation actions, without referring to a broader discussion of adaptation, while the former strongly focuses on the role of resilience, thus addressing part of the adaptation discussion only. In addition, Milestad and Darnhofer (2003) base their assessment of the adaptive capacity of organic agriculture on a comparison of the IFOAM standards with the determinants of resilience, largely without reference to empirical findings.

Organic agriculture has a long tradition in utilizing locally adapted varieties and cropping practices. It is thus able to account for the regional and local variations in climate change impacts by choosing locally adapted solutions. In addition, crop diversity is traditionally very high on organic farms. This reduces the risk of total production losses due to climate change impacts such as pest and disease outbreaks, as those impacts tend to affect different crops differently. Enhanced biodiversity also tends to reduce pest outbreaks and the severity of plant and animal diseases (Smith et al., 2011; for more details, see Niggli, 2009) and complex crop rotations break the life cycle of pests. Higher biodiversity also leads to optimal water and nutrient use as crops with different needs can be combined (in particular in polycultures).

High biodiversity generally increases the resilience of agro-ecosystems (Altieri and Nicholls, 2006; Campbell et al., 2009). Due to lower inputs in organic farming and correspondingly lower upfront costs, the economic risk of crop failures is reduced as well. The potential to realize an organic premium in certified organic markets adds to the possibility to improve the economic situation. These aspects function as an inexpensive but effective insurance against crop failure or yield reduction, and indebtedness (Scialabba and Hattam, 2002; Eyhorn, 2007). Organic agriculture has thus promising potential as a risk-reduction and risk-management strategy in the face of climate change.

Soil organic matter is increased and stabilized by farming practices common in organic agriculture (mainly compost and manure use, higher crop diversity in crop rotations and on a farm level and crop rotations with grass-clover [i.e. legumes] leys). This leads to an on-average higher water holding and retention capacity of soils under organic management. Organic agriculture is thus less vulnerable to drought, flooding and waterlogging, and erosion (Niggli, 2009; for some further details, Scialabba and Müller-Lindenlauf, 2010; Muller et al., 2011). In addition, soils are usually covered with mulch or cover crops and bare fallows are avoided. Key impacts of climate change, namely increased frequency of extreme weather events, erosion and increased water stress and drought are thus optimally addressed in organic agriculture. This property of organic agriculture is not necessarily a systemic effect, but mainly the consequence of certain common organic practices, such as the use of organic fertilizer and optimized crop rotations. These practices also increase soil quality and fertility as measured in soil nutrient availability, soil structure and aeration, and the biological diversity of soil microbes, insects and earthworms. This focus on soil quality, health and fertility also increases plant health and thus reduces vulnerability towards pest and disease outbreaks and drought (Altieri et al., 2005). The absence of synthetic fertilizers also avoids soil acidification and the corresponding soil quality deterioration (Mäder et al., 2006).

All these aspects contribute to reduce the exposure, sensitivity and vulnerability of the adapting entities. Table 5.1 contains a list of potential adaptation actions in agriculture. Many of these are common in organic agriculture or perfectly fit this cropping system, but in principle, all these practices can be implemented in a conventional context as well. For some activities, difficulties are likely in conventional systems, though. This is so for activities that rely on a systemic understanding of the farm (e.g. for pest control through increased biodiversity), which is not well established in conventional agriculture. Further details can be found in the literature, e.g. Milestad and Darnhofer (2003), Osbahr et al. (2010), Borron (2006), Howden et al. (2007), Niggli (2009), Campbell et al. (2009) or Boomiraj (2010). It is interesting to note that the full adaptation potential of fertile soils with a high organic matter content, of crop rotations and of biodiversity is hardly acknowledged in the conventional adaptation discussion (see Table 5.1).

Performance of organic agriculture in the face of the specific challenges to adaptation

a) In the case of the formalized organic sector (certified organic value chains), the aspect that the scale of adaptation measures should fit the scale of the corresponding perturbation can pose problems to organic agriculture, where the price premium often plays a decisive role and transaction costs and power inequalities along the value chains can pose considerable risks (see Chapter 6). The price premium and certified value chains for organic products in the South, however, rely often on the markets in the North, as export orientation is a common characteristic of organic production in the South. This dependence increases vulnerability, and although adopting organic production has many adaptation benefits, this aspect can have adverse effects, as additional different market risks become important. Establishing local markets and structures that reduce transaction costs and lowering the dependence on the price premium (e.g. due to lower subsidies for conventional production and its inputs or due to payments for ecosystem services) could remedy this problem (see Chapter 6).

Table 5.1 Adaptation measures in agriculture. General suggestions from the literature, indication whether they are common practice in organic agriculture (OA) and whether they have been suggested in literature focusing on organic agriculture only but not in the literature on conventional agriculture (CA). In addition, it is indicated whether the adaptation measures have some mitigation potential (the common mitigation potential due to higher yields of adapted crops and management in contrast to non-adapted ones is not indicated).

Key impact category and measures suggested in the literature	The measure is:			Remarks
	common practice in OA	not proposed in CA	also mitigating	
Regional variation				
use locally appropriate varieties and breeds	x			for this and the following two measures, high crop diversity is needed
engage in experimentation with practices, crops, varieties in the local context	x			Rosenzweig and Tubiello (2007) point out that this may take decades
engage in locally adapted breeding	(x)			
use farmer and community knowledge	x			
use traditional knowledge	(x)			
use knowledge of the ecological processes involved in agricultural production in this region	x			this can be a transformation
change land-use patterns				
Water				
change the sowing, planting and harvest dates				
change to varieties/species with vegetation and reproduction periods that fit water availability				
change to varieties/species with lower water demand		x		Rosenzweig and Tubiello (2007) provide examples that this does not always work
combine different crops such as to improve nutrient- and moisture-use efficiency due to different, non-interfering needs	x			
develop adapted cultivars	(x)			
use laser levelling to improve water management and reduce water needs				

Measure				Notes
landscape measures (shelter belts, hedges as windbreaks, shade trees, etc.) to conserve soil moisture	(x)		x	financial means are needed for this and there will be competition for water use with other sectors
invest in infrastructure for irrigation				mitigating through less energy needs for irrigation
use rainwater harvesting, water leakage reduction and water conservation techniques	(x)			time and amount of irrigation, accounting for the need of different crops (intercropping), etc.
improve water management	(x)			– improves soil structure, water infiltration, water-retention capacity and water capture and thus makes crops more resilient against water scarcity
increase soil organic matter through crop residues, manure, compost, increased diversity in rotations (including legumes), and reduced tillage (not 'no'-tillage)	x		x	– mitigating through soil carbon sequestration, less erosion (reduced soil carbon losses) and reduced energy needs (irrigation and chemical fertilizer production) this can be a transformation
change the location of cropping activities				
Pest, disease and weed pressure				
use resistant varieties	x			related to the concepts behind organic in as much as monitoring and prevention are emphasized
use integrated pest management (i.e. organic principles [monitor/prevention/control] – plus classical pesticides)	(x)			
use crop rotations (those can break the weed/pest life cycles)	x	x	x	this and the following three measures together comprise key parts of organic pest management mitigating through increased soil carbon sequestration for certain types of crop rotations that are common in organic agriculture (including legume leys, for example). With legumes, also mitigating due to reduced chemical fertilizer inputs

Table 5.1 Continued

Key impact category and measures suggested in the literature	The measure is:			Remarks
	common practice in OA	not proposed in CA	also mitigating	
plant special plants around the fields, that attract predators or repel or distract pests	x	x		
increase biodiversity on the farm and landscape level (reduces pest damage, provides habitat for predators, etc.)	x	x		
increase soil organic matter: this enhances nutrient buffer capacity and leads to higher activity of microorganism, thus increasing soil fertility and plant health	x	x	x	mitigating through soil carbon sequestration, less erosion (reduced soil carbon losses) and reduced energy needs (irrigation)
Extreme events				
improve water management (time and amount of irrigation)	x			against drought
limit livestock numbers per ha to optimal stocking rates	x			reduces overgrazing and erosion works against high temperatures needs investments
change of grazing time during the day				
build infrastructure that helps to cool stables				– provides shadow against high temperatures
agroforestry with livestock			x	– mitigating through increased soil and biomass carbon sequestration
change species/breeds for drought and temperature tolerance				
increase soil organic matter through crop residues, manure, compost, increased diversity in rotations (including legumes), and reduced tillage (not 'no'-tillage)	x			improves soil structure and fertility, water infiltration and retention capacity. Erosion, floods from heavy precipitation (in particular on a landscape scale) and water logging are thus avoided and crops become more resilient against droughts

Measure			Comment
		x	mitigating through soil carbon sequestration, less erosion (reduced soil carbon losses) and reduced energy needs (energy and chemical fertilizer production)
avoid bare fallows and use cover crops, inter crops and mulching	x		
use landscape elements such as forest shelterbelts, trees and hedgerows, etc.		x	reduce logging and makes crops more resilient against droughts – protect against winds and modifies the microclimates so that extreme temperatures can be coped with (shade, etc.) – mitigating through increased soil and biomass carbon sequestration

Variability and risk

Measure			Comment
diversify the farming system: e.g. integrate crop/livestock systems (enhances nutrient use), use crop rotations, polycropping, agroforestry, plant trees and hedgerows as habitats for other species, predators, etc.	x	x	– Campbell et al. (2009) claim that this seems promising but that empirical evidence is lacking – part of these actions mitigate through soil and biomass carbon sequestration
change time of cropping activities			
change the location of cropping activities			this can be a transformation
use short- and long-term weather forecasts			
reduce input costs	x		
develop insurance alternatives (e.g. via local financial pools)		x	reduces economic risks of crop failures
develop new financial instruments for local communities and people (e.g. micro-insurance, microfinance, etc.)			
establish support networks	x		both open and exclusive networks seem successful – for the latter, matters of justice need to be kept in mind
take on migrant work, also long term			this is a transformation
focus on stability and resilience of the farming system rather than on increasing production	x		

Table 5.1 Continued

Key impact category and measures suggested in the literature	The measure is:			Remarks
	common practice in OA	*not proposed in CA*	*also mitigating*	
Beyond the agricultural sector				no measure mentioned – this illustrates that adaptation in agriculture is still discussed with a strong sectoral focus

Sources: Compiled from Boomiraj et al., 2010, Scialabba and Müller-Lindenlauf, 2010, Niggli, 2009 (Niggli et al., 2009 contains similar information), Osbahr et al., 2010, Smith and Olesen, 2010, Campbell et al., 2009, Adger et al., 2007, Rosenzweig and Tubiello, 2007, Easterling et al., 2007 (the information used there can also be found in Howden et al., 2007), Borron, 2006.

On another level, organic agriculture is, however, well in line with the maxim of corresponding scales, as it relies on locally adapted varieties and practices that correspond to the local impacts.

b) The fact that resilience is not equal to adaptive capacity and can even be a disadvantage to adaptation where transformation would be necessary is a challenge for organic agriculture. Organic agriculture depicts itself often as a stable and resilient production system. As long as perturbations are small, this is an advantage, but awareness has to grow that transformation may become necessary and that resilience can then become a hindrance. This does not mean that resilience should not be increased but rather points to the necessity that situations where drastic changes (i.e. transformation) are needed rather than stability are recognized at an early stage.

c) The fact that certain adaptation paths may necessitate that the adapting entity disappears during the adaptation process is problematic for any farming system. It may be a particular challenge for organic agriculture, though, where agricultural production is seen as part of a living agroecosystem and not merely as a productive operation employing several natural resources, machinery and labour. Accepting that the ecosystem context shifts in such a way that supporting agricultural production is no longer feasible may pose particular difficulties. On the other hand, the tight linkage to the ecosystem context may also trigger early awareness of the need for fundamental transformation. In any case, the option of fundamental transformation must not be suppressed. The adaptation discussion in agriculture tends to neglect this and this aspect thus needs to be strengthened.

d) Equity is an important topic in adaptation projects. Different to what one may expect, organic projects are not necessarily particularly aware of equity aspects and a wide range of approaches regarding social criteria can be observed. The IFOAM guiding principles require absence of child and forced labour and freedom of association and some related issues, while further social standards (e.g. for working conditions, etc.) have only recommendatory character (IFOAM, 2005). The European Council regulation on organic production does not touch on social issues at all (EC, 2007). The Bio Suisse label, on the other hand, has detailed social regulations that go beyond the legal requirements and that make many of the IFOAM regulations into legally binding rules for Bio Suisse certification (Bio Suisse, 2011).

e) How organic agriculture performs regarding the various barriers for successful adaptation remains to be resolved on a case-by-case basis. Based on the barriers in the understanding, planning and management phases of adaptation actions as listed in Moser and Ekstrom (2010: tables 1–3), we identify only a few aspects where organic agriculture may systematically differ from conventional agriculture. These are related to information gathering and processing, where organic agriculture may have some advantage. This is because organic agriculture, as an information-intensive system and measures regarding extension services and training are usually implemented in organic projects. This thus concerns the following barriers: existence and detection of a signal, availability and accessibility of and receptivity to information, availability and accessibility of data on potential options for actions, and willingness to learn.

f)   Learning is a crucial aspect for successful adaptation. Organic agriculture has great potential to perform well regarding this challenge, as organic agriculture has a well-established learning and extension context, not the least due to the high information requirements during the conversion phase and the early years of operation. The lack of learning, and access to learning tools, etc. is thus remedied in organic agriculture and adaptation can be an easy add-on. In particular, OA is well positioned to utilize local and indigenous farmer knowledge and adaptive learning which is seen as an important source for adaptation in farming communities (Tengö and Belfrage, 2004; Salinger et al., 2005; Stigter et al., 2005; Nyong et al., 2007). Besides providing an optimal learning context, learning needs in OA are partly reduced, as many practices that are optimal for adaptation are already present in organic agriculture.

The role of learning in OA is also exemplified by Morgan and Murdoch (2000) in their stylized comparison of conventional and organic food chains where they find that

> the conventional chain is biased towards standardised knowledge with the effect that tacit knowledge is debased so that it cannot easily be drawn upon once this chain moves into crisis. In contrast, the organic model affords more scope for the utilisation of tacit knowledge in combination with benign standardised forms. This combination aims to revalue local knowledge, local ecosystems and local identities so that farmers can once again become 'knowing agents', able to exercise more autonomy and control over both their relations with other actors in the food chain and means of production on the farm.
>
> (Morgan and Murdoch, 2000: 171)

This favourable learning context in organic agriculture can also be utilized to provide the learning spaces that are important for anticipatory learning for adaptation, as e.g. promoted in Tschakert and Dietrich (2010). Thereby, external facilitators – as often present in organic contexts – can play an important role. Equity issues and power relations have to be closely monitored, though, to avoid adverse developments.

g)   Planning, implementation and evaluation for adaptation are tightly linked to learning. These are challenging tasks, in particular as they have to be kept flexible in order to react to unforeseen changes, respondent as different types of measures can have different costs under wrong forecasts (no-, low- and high-regret measures, cf. World Bank, 2010: guidance note 6). A static approach has to be avoided. A promising frame for this is adaptive (co)management, which is based on 'emphasizing the importance of understanding feedback from the environment and systematic (i.e., non-random) experimentation in shaping future actions (Berkes and Folke, 1998). It uses management as a tool not only to change the system, but as a tool to learn about the system' (Milestad and Darnhofer, 2003: 87; see also Folke et al., 2002; Olsson et al., 2004). Adaptive (co)management is thus tightly linked to learning. Adaptive management perfectly fits to organic production systems, due to their focus on ecosystem and landscape levels and their services, which necessitates constant learning and incorporation of feedback effects of actions.

h) Finally, systemic aspects need to be accounted for. This is one basis of organic agriculture and it is thus best suited to meet this challenge. Scialabba and Müller-Lindenlauf (2010), Niggli (2009), Borron (2006) and Milestad and Darnhofer (2003) all illustrate this key aspect for organic production systems. This systemic property prevents organic agriculture being a mere collection of single farming practices – each of which could in principle also be implemented in conventional agriculture. Milestad and Darnhofer (2003) point out that some organic farms go in this reductionist direction though, due to the increasing spread of industrialized food system practices to organic food. In the light of successful adaptation, moving away from systemic approaches clearly should be stopped.

Synergies between adaptation and mitigation in organic agriculture

In this section, we briefly address mitigation in organic agriculture as a livelihood strategy (see also Case study 1) and as far as it relates to adaptation. Detailed overviews on mitigation in organic agriculture can be found in Muller and Aubert (2012), Muller et al. (2011), Scialabba and Müller-Lindenlauf (2010) or Niggli et al. (2009). Mitigation can be a livelihood strategy where either synergies between mitigation and adaptation actions arise, e.g. from increased soil carbon sequestration, or it leads to additional income for farmers, e.g. from the carbon markets or via payments for ecosystem services.

Synergies between mitigation and adaptation measures

Synergies exist between mitigation and adaptation measures in agriculture. While Smith and Olesen (2010) assess this rather favourably, Rosenzweig and Tubiello (2007) remain more critical. There are some clear options for synergies, however (see Table 5.1). First, successful adaptation that directly improves yields in contrast to the situation without this adaptation always has some mitigation benefit. This is so, as in such cases, assuming that emissions largely stay unchanged, yields increase under adaptation, thus reducing relative emissions per output. This is thus a common mitigation effect of all adaptation that sustains or increases yields. Second, adaptation measures that increase or sustain soil organic carbon or biomass carbon are also mitigation measures. Examples are the use of organic fertilizers, crop rotations with legume leys or agro-forestry, and measures that reduce soil erosion. Third, adaptation measures such as those including legume leys and organic fertilizers reduce the use of chemical fertilizers and corresponding emissions from their production. Finally, energy use for irrigation and corresponding emissions are reduced in measures that improve water and moisture management and reduce irrigation needs. Examples are again measures that increase soil carbon levels, but also water-harvesting techniques. As can be seen from Table 5.1, most of these synergetic adaptation measures are common practice in organic agriculture, and many of those are proposed exclusively in an organic context.

Although there is a common physical base for many mitigation and adaptation activities, they are not the same. Adaptation measures are usually incorporated in a systemic context, while mitigation measures often are addressed as single specific agricultural practices without further relation to the broader context. Thus, while adaptation measures can realize some mitigation potential without further amendments, the reverse need not hold. Framed in the livelihood context, mitigation measures can be realized while focusing on the natural, physical and human capital, while successful adaptation requires consideration of many more aspects of the livelihood framework.

Mitigation measures and additional income for farmers

Payments for mitigation services can basically accrue in three ways. First, there are direct payments in the context of carbon markets, i.e. payments for – in some way – certified emission reductions or sequestration rates. Examples are payments for carbon sequestration under one of the various methodologies available in the voluntary carbon market or crop residue-based bioenergy projects under the Clean Development Mechanism (CDM) defined in the Kyoto Protocol. Second, there are payments for other services that correlate with mitigation achievements. Examples are payments for certain ecosystem services, e.g. for biodiversity-enhancing measures such as set-aside land, which at the same time increase carbon sequestration. We also mention the bioenergy sector, where farmers can directly benefit by growing energy crops. This is, however, no mitigation action for the farmers but rather for the energy sector. The farmer just produces the input demanded by this sector. Regarding mitigation and adaptation on the farm level, it is thus no different than other crops.

All these measures can generate additional income for the farmer. As with the synergetic measures, mitigation can be addressed in a technical approach focusing on single practices, although this does not ensure adaptation. This simplistic approach is also supported by the current institutions of the carbon markets as they build on reliable quantification and monitoring of mitigation levels. This fosters standardized agricultural production environments where comparability to a business-as-usual baseline against which mitigation levels are to be calculated is easily possible.

Payments for ecosystem services have better potential to provide mitigation while supporting adaptation, as they can be built on systemic aspects of the agricultural production system (e.g. measures for erosion control or enhancing biodiversity). Payments are executed for ecosystem services and not for mitigation. Therefore, no monitoring of the mitigation levels and comparison to a baseline is needed. The drawback of this clearly is that the mitigation achieved can only be calculated indirectly (e.g. from default values for sequestration under certain types of land use).

Conclusions

The discussion in this chapter shows that organic farming systems hold a strong potential for adaptation to climate change due to the holistic and integrated approach that minimizes vulnerability to the effects of climate change.

First, this assessment shows that organic agriculture is perfectly in line with many of the core principles and priority themes for adaptation in agriculture as formulated by well-established institutions such as the FAO (2011), for example: It exhibits synergies between mitigation and adaptation, relies on ecosystem and long-term approaches, and on location-specific adaptation activities and diversification, and it is perfectly compatible with a focus on food security (cf. also Chapter 2) and with mainstreaming climate change into development. Also the World Bank emphasizes the focus on local communities and the importance of environmental services for adaptation as two core principles (World Bank, 2010: introduction). Organic agriculture also deals well with many of the key challenges for adaptation in agriculture (barriers, learning, planning and systemic aspects). It clearly has to be cautious regarding others (scale, resilience, disappearance of the adapting entity and equity), but not more so than conventional agriculture.

Second, organic agriculture provides a context for most of the concrete adaptation measures suggested in the literature. Organic agriculture is particularly advantageous in measures for increasing soil organic matter, which is key for plant health, water management and resilience against extreme events, and in measures for increasing biodiversity, which is key in risk reduction and as a strategy against the increasing pest and weed pressure.

Third, we have found that the current adaptation discussion in agriculture does not acknowledge the full potential of soil fertility, crop rotations and biodiversity. These issues are not neglected, but they are not assessed on a systemic level, where they develop their full potential. In particular their role for plant health and pest and disease management is not developed in due detail. This lack needs to be remedied to harvest the full adaptation potential of these measures in agriculture. Acknowledging this in governmental bodies and consulting agencies, adapting policies accordingly and providing information and training on this are first steps to achieve this.

Fourth, we point out that climate change mitigation tends to relate to single practices and not to the whole organic system, while adaptation has strong systemic aspects, which are better captured by a systemic approach such as organic agriculture. Mitigation can be approached on a purely technical level, while adaptation needs a much more encompassing approach as for example outlined in the livelihood framework. Successful adaptation will often also realize some mitigation goals. This is the case for adaptation measures related to soil health, for example. Successful mitigation, on the other hand, will lead to successful adaptation only in combination with additional efforts. This has to be kept in mind when aiming at harvesting synergies between mitigation and adaptation.

Finally, special emphasis may be given to the fact that transformation and aspects beyond the agricultural production sector tend to be neglected in the discussion on adaptation in agriculture, both in organic and conventional contexts. For this, reliable long-term forecasts and planning are key. If transformation becomes unavoidable, early action needs to be taken to first implement adaptation measures that increase resilience to gain as much time as needed, but also to develop alternative livelihoods, such as identifying alternative sources of income or viable target regions for migration, and corresponding development plans for these regions to secure livelihoods for the migrants.

References

Adger, N., Arnell, N. and Tompkins, E. (2005) 'Successful adaptation to climate change across scales', *Global Environmental Change*, 15, pp. 77–86

Adger, W.N., Agrawala, S., Mirza, M.M.Q., Conde, C., O'Brien, K., Pulhin, J. et al. (2007) 'Assessment of adaptation practices, options, constraints and capacity', in M.L. Parry, O.F. Canziani, J.P. Palutikof, P.J. van der Linden and C.E. Hanson (eds), *Climate Change 2007: Impacts, Adaptation and Vulnerability*. Contribution of Working Group II to the Fourth Assessment Report of the Intergovernmental Panel on Climate Change, Cambridge University Press, Cambridge, UK, pp. 717–743

Altieri, M. and Nicholls, C. (2006) *Agroecology and the Search for a Truly Sustainable Agriculture*, Berkeley, University of California

Altieri, M., Ponti. L. and Nicholls, C. (2005) 'Enhanced pest management through soil health: toward a belowground habitat management strategy', *Biodynamics* (Summer), pp. 33–40

Bio Suisse (2011) 'Bio Suisse Standards for the production, processing and marketing of produce from organic farming', edition of 1 January, http://www.bio-suisse.ch/media/en/pdf2011/rl_2011_e.pdf

Boomiraj, K., Wani, S., Garg, K., Aggarwal, P. and Palanisami, K. (2010) 'Climate change adaptation strategies for agro-ecosystems – a review', *Journal of Agrometeorology*, 12, 2, pp. 145–160

Borron, S. (2006) 'Building resilience for an unpredictable future: how organic agriculture can help farmers adapt to climate change', Food and Agriculture Organization of the United Nations (FAO)

Burton, I. and Lim, B. (2005) 'Achieving adequate adaptation in agriculture', *Climatic Change*, 70, pp. 191–200

Campbell, A., Kapos, V., Scharlemann, J.P.W., Bubb, P., Chenery, A., Coad, L., et al.

(2009) 'Review of the literature on the links between biodiversity and climate change: impacts, adaptation and mitigation', Secretariat of the Convention on Biological Diversity, Montreal, Technical Series No. 42

Carlsson, C. and Berkes, F. (2005) 'Co-management: concepts and methodological implications', *Journal of Environmental Management*, 75: pp. 65–76

Cumming, G., Cumming, D. and Redman, C. (2006) 'Scale mismatch in social-ecological systems: causes, consequences, and solutions', *Ecology and Society*, 1, 1, article 14

Davidson, E. and Janssens, I. (2006) 'Temperature sensitivity of soil carbon decomposition and feedbacks to climate change', *Nature*, 440, pp. 165–173

Easterling, W.E., Aggarwal, P.K., Batima, P., Brander, K.M., Erda, L., Howden, S.M., et al. (2007) 'Food, fibre and forest products', in M.L. Parry, O.F. Canziani, J.P. Palutikof, P.J. van der Linden and C.E. Hanson (eds), *Climate Change 2007: Impacts, Adaptation and Vulnerability*. Contribution of Working Group II to the Fourth Assessment Report of the Intergovernmental Panel on Climate Change, Cambridge University Press, Cambridge, UK, pp. 273–313

EC (2007) 'Council Regulation No 834/2007 of 28 June 2007 on organic production and labelling of organic products and repealing Regulation (EEC) No 2092/91', *Official Journal of the European Union*, L 189/1, pp. 1–23

Eyhorn, F. (2007) 'Organic farming for sustainable livelihoods in developing countries: the case of cotton in India', vdf-Hochschulverlag, Zürich, http//www.zb.unibe.ch/download/eldiss/06eyhorn_f.pdf

FAO (2011) 'FAO-ADAPT: Framework programme on climate change and adaptation', Food and Agriculture Organization of the United Nations, FAO, www.fao.org/docrep/014/i2316e/i2316e00.pdf

Folke, C., Carpenter, S., Elmqvist, T., Gunderson, L., Holling, C. and Walker, B. (2002) 'Resilience and sustainable development: building adaptive capacity in a world of transformations', *Ambio*, 31, 5, pp. 437–440

Fuessel, H.M. (2007) 'Adaptation planning for climate change: concepts, assessment approaches, and key lessons', *Sustainability Science*, 2, pp. 265–275

Howden, S., Soussana, J.-F., Tubiello, F., Chhetri, N., Dunlop, M. and Meinke, H. (2007) 'Adapting agriculture to climate change', *Proceedings of the National Academy of Sciences of the United States of America*, 104, 59, pp. 19691–19696

IFOAM (2005) 'Social Justice', IFOAM Basic Standards, Chapter 9, International Federation of Organic Agriculture Movements, Bonn, www.ifoam.org/about_ifoam/standards/norms.html

IPCC (2007) 'Glossary', in M.L. Parry, O.F. Canziani, J.P. Palutikof, P.J. van der Linden and C.E. Hanson (eds), *Climate Change 2007: Impacts, Adaptation and Vulnerability*. Contribution of Working Group II to the Fourth Assessment Report of the Intergovernmental Panel on Climate Change', Cambridge University Press, Cambridge, UK, Appendix.

Mäder, P., Fließbach, A., Dubois, D., Gunst, L., Jossi, W., Widmer, F., et al. (2006) 'The DOK experiment (Switzerland)', in J. Raupp, C. Pekrun, M. Oltmanns and U. Köpke (eds), *ISOFAR Scientific Series*. University of Bonn, Germany

Meehl, G.A., Stocker, T.F., Collins, W.D., Friedlingstein, P., Gaye, A.T., Gregory, J.M., et al. (2007) 'Global Climate Projections', in S. Solomon, D. Qin, M. Manning, Z. Chen, M. Marquis, K.B. Averyt, et al. (eds), *Climate Change 2007: The Physical Science Basis*. Contribution of Working Group I to the Fourth Assessment Report of the Intergovernmental Panel on Climate Change, Cambridge University Press, Cambridge, UK

Milestad, R. and Darnhofer, I. (2003) 'Building farm resilience: the prospects and challenges of organic farming', *Journal of Sustainable Agriculture*, 22, 3, pp. 81–97

Morgan, K. and Murdoch, J. (2000) 'Organic vs. conventional agriculture: knowledge, power and innovation in the food chain', *Geoforum*, 31, pp. 159–173

Moser, S. and Ekstrom, J. (2010) 'A framework to diagnose barriers to climate change adaptation', *Proceedings of the National Academy of Sciences of the United States of America*, 107, 51, pp. 22026–22031.

Muller, A. and Aubert, C. (2012) 'The potential of organic agriculture to mitigate the impact of agriculture on global warming – a review', in S. Penvern et al. (eds), *Organic Farming, Prototype for Sustainable Agri-cultures?* Dordrecht, Springer

Muller, A., Jawtusch, J. and Gattinger, A. (2011) 'Mitigating greenhouse gases in agriculture – a challenge and opportunity for agricultural policies', Report commissioned by 'Brot für die Welt' (Germany), 'Brot für alle' (Switzerland), Stuttgart, DanChurchAid (Denmark) and Church of Sweden.

Nelson, D., Adger, N. and Brown, K. (2007) 'Adaptation to environmental change: contributions of a resilience framework', *Annual Review of Environment and Resources*, 32, pp. 295–419

Niggli, U. (2009) 'Organic agriculture: a productive means of low-carbon and high biodiversity food production', *Trade and Environment Review 2009/2010*, UNCTAD, pp. 112–142

Niggli, U., Fließbach, A., Hepperly, P. and Scialabba, N. (2009) 'Low greenhouse gas agriculture: mitigation and adaptation potential of sustainable farming systems'. Food and Agriculture Organization of the United Nations (FAO) April, Rev. 2, Rome, ftp://ftp.fao.org/docrep/fao/010/ai781e/ai781e00.pdf [16.02.2011]

Nyong, A., Adesina, F. and Osman Elasha, B. (2007) 'The value of indigenous knowledge in climate change mitigation and adaptation strategies in the African Sahel', *Mitigation and Adaptation Strategies for Global Change*, 12, pp. 787–797

Olsson, P., Folke, C. and Berkes, F. (2004) 'adaptive comanagement for building resilience in social-ecological systems', *Environmental Management*, 34, 1, pp. 75–90

Osbahr, H., Twyman, C., Adger, N. and Thomas, D. (2010) 'Evaluating successful livelihood adaptation to climate variability and change in Southern Africa', *Ecology and Society*, 15, 2, p. 27

Risbey, J., Kandlikar, M., Dowlatabadi, H. and Graetz, D. (1999) 'Scale, context, and decision making in agricultural adaptation to climate variability and change', *Mitigation and Adaptation Strategies for Global Change*, 4, pp. 137–165

Rosenzweig, C. and Tubiello, F. (2007) 'Adaptation and mitigation strategies in agriculture: an analysis of potential synergies', *Mitigation and Adaptation Strategies to Global Change*, 12, pp. 855–873

Salinger, M., Sivakumar, M. and Motha, R. (2005) 'Reducing vulnerability of agriculture and forestry to climate variability and change: workshop summary and recommendations', *Climatic Change*, 70, 1–2, pp. 341–362

Scialabba, N. and Hattam, C. (eds) (2002) 'Organic agriculture, environment and food security', Food and Agriculture Organization of the United Nations (FAO), Rome, www.fao.org/docrep/005/y4137e/ y4137e00.htm

Scialabba, N. and Müller-Lindenlauf, M. (2010) 'Organic agriculture and climate change', *Renewable Agriculture and Food Systems*, 25, 2, pp. 158–169

Smit, B. and Wandel, J. (2006) 'Adaptation, adaptive capacity and vulnerability', *Global Environmental Change*, 16, pp. 282–292

Smith, J., Wolfe, M., Woodward, L., Pearce, B. and Lampkin, N. (2011) 'Organic farming and biodiversity', A review of the literature. Aberystwyth, Wales

Smith, P. and Olesen, J.E. (2010) 'Synergies between mitigation of, and adaptation to, climate change in agriculture', *Journal of Agricultural Science*, 148, pp. 543–552

Stigter, C., Dawei, Z., Onyewotu, L. and Xurong, M. (2005) 'Using traditional methods and indigenous technologies for coping with climate variability', *Climatic Change*, 70, 1–2, pp. 255–271

Tengö, M. and, Belfrage, K. (2004) 'Local management practices for dealing with change and uncertainty: a cross-scale comparison of cases in Sweden and Tanzania', *Ecology and Society*, 9, 3, www.ecologyandsociety.org/vol9/iss3/art4

Tschakert, P. and Dietrich, K.A. (2010) 'Anticipatory learning for climate change adaptation and resilience', *Ecology and Society*, 15, 2, p. 11

Walker, B., Holling, C.S., Carpenter, S.R. and Kinzig, A. (2004), 'Resilience, adaptability and transformability in social–ecological systems', *Ecology and Society*, 9, 2, p. 5

Walker, B.H., Gunderson, L.H., Kinzig, A.P., Folke, C., Carpenter, S.R. and Schultz, L. (2006) 'A handful of heuristics and some propositions for understanding resilience in social-ecological systems', *Ecology and Society*, 11, 1, p. 13

World Bank (2010) 'Mainstreaming adaptation to climate change in agriculture and natural resources management projects', World Bank, http://climatechange.world bank.org/climatechange/content/mainstreaming-adaptation-climate-change-agriculture-and-natural-resources-management-project

Case study 3
Carbon credits from organic agriculture

Experiences from Costa Rica

Jonathan Castro and Manuel Amador

During 2004, Dutch donors, by means of the Humanist Institute for Development Cooperation (HIVOS), charged the Educational Corporation for Costa Rican Development (Corporación Educativa para el Desarrollo Costarricense, CEDECO), a non-profit NGO, with developing and implementing a methodology to validate the potential of small-farmer agro-ecological practices in carbon sequestration, GHG emissions reductions and energy efficiency. Based on this, carbon credits for the voluntary carbon market, on principles of direct partnership with private companies, can be generated. This case study describes the methodological approach used and the livelihood effects of this project. The University of Costa Rica, the National University and the Center for Tropical Agricultural Research and Teaching (CATIE) in Costa Rica and the Research Institute of Organic Agriculture (FiBL) from Switzerland participated in the research. The National Institute for Agricultural Sciences (INCA) in Cuba was also invited to carry out similar studies on the island.

The research

For the last seven years, CEDECO has carried out studies on the energy efficiency and mitigation potential of small-scale ecological agriculture. The goal has been to gather scientific evidence to decide whether small-scale organic growers can be included in mitigation strategies and to create new mechanisms that recognize the environmental services provided by this important agricultural sector in food production and environmental preservation.

The research was carried out in three stages. First, between 2004 and 2005 a methodology was developed to evaluate the mitigation potential of organic farms. The second stage (2006–2007) included observations on physical, chemical and biological variables from the farms and incorporated social and economic assessments proposed by CEDECO as an added value for the study. An analytical model was created that correlates the variables measured on-farm to GHG emission reductions, energy efficiency and social benefits for the peasant families.

With the technical and methodological information validated during the first two stages available, the third phase was started in 2008 (still ongoing) to ascertain the regions or farms that generate environmental services. The methodology was used to classify the organic farms from a region according to their energy efficiency. The results from this process have served as evidence for proposing the Cam(Bio)2 standard and seal.

Methodology

CEDECO developed its methodology following the IPCC Guidelines for National GHG Inventories[1] and existing methodologies from the Clean Development Mechanism and the Voluntary Carbon Market. The concrete design of the methodology was chosen in such a way as to be feasible also with the local limitations related to instrumental and analysis facilities.

The methodology was applied in phase 1 (2003–05) to farms at different altitudes and regions and on different crops. A total of 45 farms were evaluated throughout the country, dedicated to crops such as coffee, sugar cane, vegetables, pineapple and cocoa, with banana as agro-forestry systems. Farms with conventional and organic management were evaluated in the same regions, with the same crops.

The study defined priority analyses regarding soil gas emissions, energy efficiency on organic farms, organic carbon content of the soil, and soil microbiology, as well as analyses of social and economic factors. In order to analyse organic soil carbon deposits, the reference was the depth of the arable layer where determination was possible, or the first 30 centimetres of the profile, accepted for soils with agricultural crops. The samples were evaluated for variables such as organic carbon content, true horizon depth and soil density.

According to the IPCC Guidelines for National GHG Inventories, it was assumed that 1.25 per cent of the nitrogen inputs in agricultural systems are released as N_2O; therefore, for every 100 kg of N/hectare, the system will emit 1.25 kg of N (as the work was done before 2006, we used IPCC 1996 guidelines). On the basis of an analysis of the technical logs, it was possible to define a temporal space related to the crop cycles being studied and a sum of nitrogen inputs was quantified. Based on the specified area, a soil gas emission rate was estimated in relation to the application of natural and chemical fertilizers and manure.

In order to evaluate energy performance for each farm, key crops (coffee, cocoa, lettuce, coriander, sugar cane) were taken into consideration. All tasks were described and all inputs employed were quantified for each of these crops, including labour. Furthermore, crop productivity data were gathered. To effect the calculations, agricultural and animal product energy equivalents were used, as well as the energy expenditure for each input as reported in the literature. The information from each farm was processed under the program Energía version 3.0, initially developed by Sosa and Funes-Monzote (1998)[2] and updated by Funes-Monzote et al. (2009).[3] Finally, energy demands of the crop cycles (i.e. one year for coffee, three months for lettuce, etc.), energy yield from the harvest produced and its energy efficiency were determined for each region, farm and crop.

General trends

The technical evaluations carried out in Costa Rica have determined the soil carbon reserves for the different production systems. No large differences were found between organic and conventional management, but there were both maintenance and increase in these reserves associated with the application of organic fertilizers, green ground cover, agro-forestry systems and soil conservation efforts.

Differences were found between organic and conventional systems with regard to the reduction in soil emissions and increased energy efficiency in all of the crops analysed. As a result of these studies, CEDECO currently has a dossier on the contribution of organic agriculture to the mitigation of climate change through carbon sequestration, emissions reduction and improvements in energy performance.[4]

Standard and seal as a proposal for instrumentalizing the contribution of agro-ecology to climate change mitigation

The results from the study acted as a basis for drafting a third-party auditable standard for climate-friendly agricultural production. The standard seeks certification by a third party and the seal identifies products from the farms that reduce GHG emissions.

This stage has been important for operationalizing the standard and seal, while at the same time incorporating them into current recognition paths for environmental services derived from organic agriculture. The field studies permitted a statistical validation of the analytical model for the farms and the creation of a standard that would certify by means of a seal that backs this mitigation service. Support was garnered for this from the BCS Öko-Garantie certifying agency, which is headquartered in Germany.

Support is currently being sought from different quarters, including civil society, entrepreneurial initiatives responsible for emissions, and the voluntary carbon market, to accept Cam(Bio)2 as a high-quality standard for mitigation services in agriculture. These resources can be mobilized towards growers' organizations where the mitigation potential is realized.

Can small ecological farmers negotiate their environmental services?

A key point when developing such a scheme for smallholders is the overall sustainability performance besides mitigation, as e.g. assessed along the various dimensions of the livelihood framework.

Developing carbon credits for organic farming and smallholders faces considerable criticism from a broad range of institutions. Some see a danger in exposing smallholder farmers and organic agriculture to the harsh reality of the carbon markets. Others consider it unethical that the practices of organic farmers receive an economic reward, even though this type of agriculture also requires investments. This has been a complicated topic that also has limited the transformation from conventional agriculture to agro-ecological practices. Larger investments are currently needed to sustain groups of farmers that are seeking to join the market with good quality products.

CEDECO, through the credits generated in the Cam(Bio)2 project, places a value on the environmental service provided by the farmers (by lowering emissions in these communities, efficiently using energy and sequestering carbon), to create alliances with investors wishing to offset their emissions, and also to show their solidarity, and who are interested in a direct relationship with the farmers' organizations. By means of Cam(Bio)2 an option of 'high-quality' credits in the voluntary carbon market is developed. The ethical relationship between those providing the environmental service and the investors is regulated with a quantification of emissions at the point of origin, a commitment to reduction and the later compensation to the organic farmers' organizations and linked to fair trade. It is a relationship where the investment establishes an alliance with the organizations that elaborate products with an added value, and procures better income for the sustainability of the organic farmers' organizations. The standard thus quantifies reductions that are sold to companies on an individual business contact level between company and farmer group, and the credits are in this way not freely available on the carbon market. CEDECO thus tries to build voluntary alliances between companies and farmers' organizations, not necessarily in the context of the existing carbon markets.

Environmental service, sustainability of agro-ecological processes

CEDECO can complement investments in sustainable processes in the Central American and Caribbean regions through negotiations on the voluntary carbon market. With the Cam(Bio)2 initiative, tools were created to demonstrate that the regions with a higher proportion of organic farms make a considerable contribution to mitigation of the greenhouse effect. The initiative is oriented to the quantification of the amount of carbon sequestered on the farms of an organization and that is offered to firms or sectors of civil society, such as municipalities in developed countries, interested in mitigating climate change. The resources generated from the carbon credits are not distributed among the farms, but are invested in projects to resolve production problems, e.g. a fertilizer enterprise to increase yields or for added value on what is already produced, such as packaging or improving the processing of their production.

The mechanism would be the following: the stage for assessing the indicators and applying the research model generates the values for accreditation of the mitigation services by means of the standard. The certifier, as a third party, verifies the veracity of the methodological application and validates the environmental credits. Simultaneously, the regions and their agro-ecological projects are presented to firms and civil society in Europe, the USA, etc, as potential buyers. Finally, these organizations, through the accreditation of the standard, maintain the investment in development processes, which makes the scheme sustainable. Private firms and civil society would receive a bond for this investment accrediting their collaboration in development for a specific volume of carbon credits and establishing a direct channel of collaboration with the organization and the region. Thus, the resources generated return to the organizations in the form of projects that resolve problems in the value chain, or add value to the organization, thus contributing to the sustainability of the region (Figure 1).

As a result of the project's innovation and proposal, climate change adaptation criteria are currently being incorporated. We understand that in spite of the fact that tremendous efforts are being made to mitigate climate change, its consequences will have short- and medium-term effects in the rural regions and on food production. Organic agriculture has been shown in and of itself to be capable of responding in a more adequate manner to climate fluctuations, since its productive systems have a greater capacity for adaptation. In a similar fashion, the alliances for development and project implementation by growers' organizations must be oriented to activities that contribute to promoting these agricultural systems.

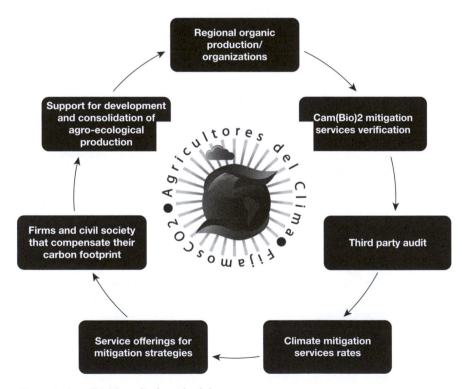

Figure 1 Cam(Bio)2 cyclical methodology

CEDECO has assumed an ethical and strategic commitment to establish regional networks where growers, support organizations and states are invited to consolidate an alternative proposal for recognizing mitigation and other environmental services from organic farmers with a Latin American perspective. The final intention is to propose new mechanisms facilitating the fight against poverty, support for rural development, mitigation of climate change and achievement of sustainable development.

The approach illustrated in this case study shows how carbon credits from ecological farming can support sustainable development in smallholder contexts. Participation in the carbon markets need not compromise sustainability – but it needs to be carefully designed to achieve this.

Notes

1   www.ipcc-nggip.iges.or.jp/public/2006gl/index.html
2   Sosa, M. and Funes-Monzote, F. (1998) 'Sistema para el análisis de la eficiencia energética de fincas integrales', Instituto de Investigación de Pastos y Forrajes. La Habana. (Software)
3   Funes-Monzote, F.R., Castro, J., Pérez, D., Rodríguez, Y., Valdés, N. and Gonçalves, A.L. (2009) Energía 3.01. Computerized system for calculating indicators of energy efficiency. CEDECO-INCA-EEPF Indio Hatuey. La Habana. (User's Manual/Software)
4   Additional information may be found at www.cambio2.org

6 Organic agriculture governance in the Global South

New opportunities for participation in agricultural development and livelihood outcomes

Kristen Lyons, Gomathy Palaniappan and Stewart Lockie

Introduction

The expansion of organic agriculture in the Global South is frequently associated with positive social, environmental and economic outcomes (see for example Parrott and Marsden, 2002; Thamaga-Chitja and Hendriks, 2008; Willer and Kilcher, 2011). Non-government organizations, international development agencies and researchers, amongst others, cite smallholders as foremost amongst its beneficiaries. In the 2000s a range of participatory organic regulatory models have emerged that aim to incorporate diverse stakeholder interests, including smallholders, into decision-making processes, thereby delivering positive outcomes for such stakeholders. The rhetoric of collaborative and inclusive decision making suggests such deliberative processes will be integral in delivering socially and environmentally sustainable agricultural development. But to what extent do emerging forms of organic governance enable actors from the Global South to actually participate in shaping the expanding organic sector? What are the impacts and outcomes of this participation – including livelihood outcomes – as well as the broader socio-ecological impacts of the expanding organic sector for the Global South?

We address these questions with particular reference to organic standards setting, inspection and certification processes, which – along with equivalence arrangements – comprise critical techniques for the governance of organic supply chains. For other types of market access and inclusion barriers, see Chapter 7. As Lockie et al. (2006) demonstrate, certification and labelling have played major roles in the constitution of the organic market. Global trends towards standardized organic certification processes involving detailed documentation and third-party verification of production practices have been driven both by the need to comply with International Organization for Standardization (ISO) requirements and to facilitate trade (Mutersbaugh and Klooster, 2010). Such standardization benefits exporters by reducing the cost of certifying to multiple national standards. However, it has also been criticized for reducing the local relevance of environmental and social criteria and of certification processes (Mutersbaugh and Klooster, 2010), for increasing the dependence of farmers on transnational certifiers (Mutersbaugh, 2004), as well as centralizing the values of Northern actors in shaping the lived realities of farmers in the South (Bacon, 2008). Such concerns, alongside the broader participatory turn in environmental and agricultural governance, have facilitated the emergence of a number of localized and participatory organic governance processes.

Our chapter begins with an overview of the expansion of organic farming and organic governance arrangements in the Global South. We critically evaluate the form of organic governance arrangements, including the emergence of Global South and smallholder-specific organic agriculture governance arrangements. Our three case studies – Uganda, India and the Philippines – demonstrate different models for including local level actors, as well as complex and nuanced outcomes in terms of participation and inclusion. Through a comparative analysis of this cross-country data, we demonstrate the impacts of participation and non-participation in organic governance arrangements upon broader livelihood and other socio-ecological outcomes. Our chapter concludes by reflecting on the broader socio-economic impacts of organics as a strategy for agricultural and rural development in the Global South.

Expansion of organic farming and organic governance in the Global South

Worldwide, the organic food and agriculture sectors have experienced sustained growth over many years, including a doubling of sales in some years since the early 1990s (Lockie et al., 2006; Willer and Kilcher, 2011). An analysis of organic sector expansion demonstrates that growth and contraction has occurred unevenly across regions and commodities, including a contraction in certain sectors, especially since the global financial crisis in 2008–2009. A notable trend includes ongoing and significant production increases in countries of the Global South, with Parrott and Marsden (2002) suggesting that approximately two-thirds of new entrants to organics are located in countries of the South.

The expansion of organics in the Global South has occurred across both: (1) the certified organic sector – including the production of organic commodities for sale on (mostly) international markets (as discussed below); and (2) the non-certified organic sector (also described as 'passive' or 'de-facto' – see Jaffee, 2007; Parrott and Marsden, 2002 respectively). Both certified and non-certified organic agriculture in the Global South, as framed by the United Nations, the Food and Agriculture Organization (FAO) and the International Federation of Organic Agriculture Movements (IFOAM), amongst others, are able to deliver multiple benefits, including food security, poverty alleviation, water security, climate change adaptation and biodiversity conservation (see for example FAO, 2007; UNEP-UNCTAD, 2008; Walaga and Houser, 2005). On the back of such claims, the International Assessment of Agricultural Science and Technology

Development (IAASTD, 2008) – drawing on over 400 of the world's leading scientists in these topics – has called for greater support for agro-ecological farming systems (including organic farming systems).

While traditional and/or subsistence farming methods – including crop rotations, intercropping, mulching and biological pest control – are frequently described as organic, it is in the certified organic sector in the South – and supported via significant international development agency investment – where recent growth has been recorded.[1] While 'de-facto' and certified organic farms may not always demonstrate significant variation in actual farming practices, certified organic farmers are required to provide *evidence* of compliance with a set of internationally recognized agronomic, ecological, animal welfare and, most recently, social criteria. Organic criteria include reference to the use of allowable inputs (e.g. animal manures and some natural herbicides), allowable practices (e.g. crop rotations and companion planting), as well as prohibited substances (including GMOs, nano particles and antibiotics). In addition, organic standards stipulate a range of social and environmental management criteria (including reference to labour relations, gender equity, biodiversity, soil fertility and water conservation), as well as detailed record-keeping requirements.

Demonstrative of certified organic sector expansion in the Global South, in 2010 an estimated 40 per cent of the world's organic producers were in Asia, followed by Africa (28 per cent) and Latin America (16 per cent) (Willer and Kilcher, 2011). The largest numbers of certified organic producers are currently found in India (677,000) and Uganda (187,893), two of the selected case studies reported on in this chapter (Kung Wai, 2011; Namuwoza and Tushemerirwe, 2011). Meanwhile the Philippines, our third case study, has the largest area under organic production across South East Asia.

The majority of certified organic produce from the Global South is sold on international – and mostly Northern – markets. The dominance of international trade of organic produce from the Global South is, in large part, an outcome of export-led development planning that dominates both conventional and organic agricultural development, and is facilitated via a range of neo-liberal policy interventions, including trade liberalization and the privatization of agri-food standards (Friedmann, 2004; McMichael, 2004; Langan, 2011). A number of development agencies, including the Swedish International Development Cooperation Agency (SIDA) and the Agro-Eco Louis Bolk Institute, describe organics as an 'export-led development strategy'; and on the basis of such claims have provided financial support to a number of export companies to assist in the costs associated with organic certification, as well as the provision of various other supports (see Lyons, 2011). While on the one hand, some local stakeholders argue the expansion of the organic export market has trickle-down effects that benefit local food systems (including the provision of organic training to farmers, increased local recognition of organic produce, etc.), at the same time, the policies and practices driving export-led development have also constrained the expansion of local (organic) production and exchange networks. In their research in Kenya, for example, Freidberg and Goldstein (2010) describe the legacies of international agricultural development – including structural adjustment programmes, trade liberalization and aid interventions – as part of the complex political terrain that constrained the longevity of a local box scheme.[2] Alternative localized food networks are left to swim against the tide of export-led agricultural development initiatives. Reflecting this struggle, the majority of African certified organic produce is exported (to the European Union, the United States and Japan among others); meanwhile Egypt and South Africa are among the few countries on the African continent with sizeable domestic markets for organic produce (Lyons, 2011 forthcoming).

South–North trade in organic produce, an outcome of the export-led development trajectory, is characterized by distant relations between producers, consumers and intermediary stakeholders. These relations are mediated by organic certification, and verified via third-party certification. Since the introduction of the first organic standard by the UK Soil Association in 1973, Willer and Kilcher (2011) estimate there are now at least 532 government and non-government agencies that offer organic certification services. The emergence of organic standards is demonstrative of the audit culture that characterizes the neo-liberal governance of food and agriculture, the results of which require agriculture to be commonly mediated by one, or a number, of quality standards (Barrett et al., 2002; Campbell and Le Heron, 2007; Jaffee, 2007; Bacon, 2008).

The introduction of organic certification systems (alongside other 'quality' standards, including Fair and Ethical Trade (see for example Smith and Lyons, 2011) into the Global South has raised a number of challenges and problems for farmers. Amongst these is the cost to farmers for ensuring compliance with international standards. Given the majority of farmers in the Global South are smallholders – who earn just a small income from their farming activities – the costs associated with certification are largely out of reach (see for example Barrett et al., 2002). In addition, farmers must comply with an array of organic (and other 'quality') criteria – and often with little explanation of the content of such criteria (see for example Lyon et al., 2010). The introduction of organic (and other third-party) certification systems has also brought with it a range of audit requirements that demonstrate a failure to consider the diverse cultural contexts in which standards are applied (see for example Dolan, 2010; Smith and Lyons, 2011). For example, in their research on audit requirements related to ethical trade and organic certification, Smith and Lyons (2011) described requirements for farmers to provide written document keeping as being inappropriate for those farmers who were illiterate. They also identified insensitivity to gendered power relations. For example, women farmers were asked to speak – in front of men – about the specific challenges for women associated with compliance with organic standards. These circumstances have been critiqued for creating new forms of South–North dependency, or 'neo-colonialism', as well as exacerbating cultural and ecological destruction (see for example Mutersbaugh, 2004).

These mounting concerns have given rise to the formation of a number of 'bottom-up', 'grassroots' and 'participatory' organic governance mechanisms.

Amongst these are smallholder and group certification schemes, as well as Participatory Guarantee Systems (PGS).

In Uganda, for example (alongside a number of other countries in the Global South), national-level actors, including the national peak body the National Organic Agricultural Movement Uganda – NOGAMU (representing various local farmer organizations), the national organic certifier Uganda Organic Certification Ltd (UgoCert), and with financial support via international donors (including SIDA) and the international peak body IFOAM, have introduced a smallholder-specific group certification system. The group certification model relies on farmers being arranged into groups, and then certified as a group (rather than individually). Group members are required to comply with criteria outlined in an Internal Control System (ICS) – the contents of which are negotiated between various local and international actors. Inspection to ensure compliance with the ICS is then delegated to local inspectors, with international certifiers undertaking random inspections of a small portion of group members on an annual basis.

In India, collaborations among a number of Indian government departments, the civil society sector and IFOAM, with initial funding from the FAO, facilitated the introduction of a Participatory Guarantee System (PGS) in 2006. The PGS provides a locally focused quality assurance scheme where the certification of producers relies on the active participation of farmers who undertake the role of local-level inspection (see below for further details). The Indian organic agriculture sector is also supported by fully implemented organic agriculture regulations and 17 national and regional certification bodies (Katto-Andrighetto, 2011).

Meanwhile in the Philippines, the majority of organic producers are certified via individual third-party accreditation, which is supported via fully implemented regulations on organic agriculture and the introduction of organic labelling regulations (Kung Wai, 2011). However in the 2000s a number of local NGOs and farming organizations have been active in establishing a smallholder group certification system, similar to the arrangements in Uganda.

This chapter examines the extent to which these emergent forms of organic governance represent *alternative* models for connecting organic producers and consumers; in ways that redress the power imbalance between South–North, and some South–South, actors (see also Friedmann and McNair, 2008). To address these themes, our chapter examines the extent to which emergent global agri-food relations, facilitated via the introduction of smallholder and participatory regulatory models, provide arenas for local actors to define culturally and ecologically region-specific understandings of 'organics'. In other words, are local-level actors able to shape organic standards, standard-setting processes and audit arrangements in ways that enable them to play a role in shaping both farming practice and rural development outcomes? While group and participatory certification schemes are frequently couched in discourses of empowerment, inclusion and partnership, such discourses can mask the power relations that shape such negotiations, and temper the extent to which organic governance arrangements represent contested sites. This chapter aims to shed light on these contested spaces, and the impacts of outcomes of this contestation for rural livelihoods.

Research methods: introduction to the case studies

The three countries selected for analysis in this chapter are among a number of countries in the Global South that have experienced significant recent organic sector expansion. Uganda currently comprises the largest area under certified organic production in Africa, with an estimated 226,954 hectares certified organic, a figure that represents around 1.74 per cent of Uganda's total agricultural land (Willer and Kilcher, 2011). Uganda has also been active in establishing national and regional organic governance arrangements, including the registration of a national certifier 'UgoCert' in 2004, and – along with other East African countries – the East African Organic Product Standard (EAOPS) and mark (Kilimohai) in 2007. India is the second case study included in this chapter. India currently comprises the largest number of organic producers in the world, an estimated 677,000 certified organic farmers (Willer and Kilcher, 2011). This figure has increased significantly since the year 2000; circumstances which in part explain the rapid expansion in global figures. There is currently an estimated 1,180,000 hectares certified organic farmland in India, an area that comprises 0.66 per cent of India's total agricultural land. In the Philippines, the third case study included in this chapter, 52,546 hectares of land are officially recognized as certified organic, an area that represents 0.45 per cent of Philippines' agricultural land. A recent pledge by the Department of Agriculture of 900 million Philippine pesos (approximately 15 million euros) is expected to boost the country's organic agriculture programme over the coming years (Organic World Net, 2010).

The findings presented in this chapter draw from qualitative and quantitative research conducted by the authors between 2005 and 2009 in Uganda (Lyons), 2007 in India (Palaniappan) and between 2007 and 2008 in the Philippines (Lockie).

In Uganda, fieldwork by Lyons included in-depth face-to-face interviews with 30 smallholder organic farmers certified via group certification with international organic certifiers (including IMO and KRAV) and via both international and domestic (UgoCert) inspectors. Many of these farmers were members of the Katuulo Organic Pineapple Cooperative Society (KOPCS), located in the Kyazanga Sub County, south west of Kampala, as well as coffee growers at Sipi Falls in the Kapchorwa district who supplied Kawacom International, an international coffee trader. Interviews were also conducted with representatives from Amfri Farms (a horticultural export company – the buyer of KOPCS produce), Kawacom Uganda Ltd as well as NGO and international development agency representatives in Uganda. Qualitative data were analysed via thematic coding.

In India, Palaniappan[3] conducted convergent face-to-face interviews along with a co-researcher, native to the area, to coordinate the interviews. Eighteen smallholder organic farmers were interviewed who were converting to organic farming through a farmers' network in Tamil Nadu. A heterogeneous sample of participants (across gender and age categories) was chosen through different sources, including snowball sampling via key gatekeepers, as well as via magazines and websites. The interviews were taped and the information was analysed both during the interviewing process, and after transcribing the recordings.

Philippine data are drawn from a detailed livelihoods survey conducted in 2007–2008 among 347 farming households from six *barangays* (villages) in the uplands of Negros Occidental (Lockie et al., 2010). Following the sustainable livelihoods framework (DFID, 2001), data were collected on household structure and demographics, livelihood assets, livelihood activities, involvement in development and conservation activities, and livelihood outcomes. These were supplemented with pre- and post-survey interviews with *barangay* leaders and groups of farmers.

Organic agriculture in Uganda – constrained possibilities for participation

The first certified organic agriculture projects began in Uganda in the early 1990s, most of which provided produce for international markets. Organic produce and products include organic solar-dried tropical fruits, chillies, ginger, beans and okra, and an organic cotton export initiative, as well as vanilla, bark cloth, cocoa, Arabica and Robusta coffee and sesame. Until recently, organic certification was mostly undertaken by international organic certifiers, including IMO, EcoCert, BCS and KRAV. In recent years, the certification organization UgoCert has registered to provide certification and inspection services. Ugo-Cert has devised a domestic organic standard, as well as introducing locally specific audit and inspection requirements – including the introduction of group certification.

Group certification

This auditing arrangement is based upon the organization of smallholder farmers into groups, and with an organic certificate awarded to the group rather than to individual farmers. The buyer of the organic produce – including the coffee trader, Kawacom International, and horticultural exporter, Amfri Farms – covers payment for group certification. Under this arrangement, recognition of smallholder farmers' organic status is tied to their relationship with a buyer. Management of the group occurs via an Internal Control System (ICS); an internal quality control document that stipulates requirements related to growing methods, post-harvest handling, record keeping and other activities (see also Lockie et al., 2006). Inspection is then undertaken by local inspectors with an additional annual inspection of a percentage of smallholders by an international external evaluator. The group model has provided a new mechanism for small-holders to obtain organic certification from international organizations, thereby enabling smallholders to participate in (mostly) international certified organic markets.

While group certification has reportedly occurred since the 1980s, it has only been since the year 2000 that the organic sector has sought to clarify terminology and harmonize principles that underpin these arrangements. Between 2001 and 2003, IFOAM ran a series of workshops so as to reach consensus on a range of issues related to group certification, including: acceptable definitions of 'small-holder' farmers, minimum requirements for the ICS, inspection processes and rates of re-inspection, as well as processes to manage non-compliance (van Elzakker and Rieks, 2003). While the outcomes of these workshops have helped achieve international harmonization on ICS, there are still limits to international recognition of group certification schemes (most notably in the United States and Japan).

The introduction of group certification has delivered new opportunities for some smallholders in Uganda. Some smallholder organic coffee producers at Sipi Falls, for example, discussed membership in the organic group as providing an assured buyer for their organic coffee (Kawacom International), which covered the costs associated with certification. Kawacom also provided a range of resources to support farmers to comply with the ICS (including tarpaulins for drying coffee beans) and organic farm training (including information on cover crops, crops for use as natural insecticides and terracing to prevent soil erosion). Some smallholders also discussed the benefits of receiving immediate payment from the sale of their coffee, their close proximity to an organic buyer at the Sipi Falls trading centre, as well as the issuing of receipts as assisting in monitoring crop harvests.

At the same time, however, some smallholders also expressed frustration at the limited quantity of coffee beans they were able to sell as organic to Kawacom; with most interviewees indicating they sold just a small portion of their total organic crop. This was reiterated by a representative from NOGAMU, who stated that supply significantly outstripped demand across the entire Ugandan organic sector, including the organic coffee sector. While the challenge of finding markets for organically produced coffee (and other products) is not unique to smallholder farmers, the group certification model presents specific issues. For example, smallholder coffee producers described their experience of 'competing' with other group members for the sale of their coffee, and in such circumstances, expressed frustration about the lack of transparency by which their buyer selected coffee from certain farmers over others. Some smallholders despaired that buying arrangements were shaped by nepotism, with local inspectors and buyers favouring family members and friends above other group members. Similarly, smallholders expressed concerns that the two coffee buying seasons left them without an income from the sale of organic coffee for many months of each year. Some smallholders also expressed frustration at the limited support Kawacom had provided to assist in establishing the organic coffee industry at Sipi Falls. For example, a few coffee growers lamented that early supports – including the provision of tarpaulins and farming tools, as well as catering for group meetings – had dwindled away. Despite these frustrations, smallholders' organic status was tied to their group certificate, circumstances which left smallholders with little option but to sell organic coffee to Kawacom.

Smallholders from the KOPCS – who had group certification that was financed by Amfri Farms – articulated similar frustrations. KOPCS members lamented that while Amfri Farms had regularly purchased organic pineapples from their cooperative over many years, the majority of their organic produce was sold – unlabelled and without a price premium – on the local market. Smallholders were aware they competed not only with other group members from their cooperative for the sale of their produce, but with smallholders from other regions, including regions closer to the Amfri Farms' processing house in Kampala.

Local inspection

The introduction of localized organic inspection arrangements – often as part of group certification – has also delivered mixed opportunities and outcomes for smallholders in Uganda. In 2004, UgoCert (alongside NOGAMU, export companies and other actors) was successful in advocating for local inspectors to undertake organic inspection on behalf of international certifiers. These actors reported a number of benefits associated with the introduction of local inspection, including a reduction in the cost of inspection, year-round availability of organic inspectors, and the increased likelihood that inspection processes and inspectors were cognizant of local social, cultural and ecological contexts. A representative from UgoCert explained that while these aspects of local inspection had delivered benefits for smallholders, they had also delivered benefits for export companies (including Kawacom International, Amfri Farms, as well as the growing number of organic export companies active in Uganda – counted at 42 in 2009–2010 (Namuwoza 2011)). As such, exporters were foremost amongst those lobbying for international recognition of local inspectors:

> Many operators here in Uganda are actually putting pressure on their certifiers to use the local inspectors, so that the costs can be reduced. So, many of the operators are refusing to meet the costs of flying in an inspector from the UK, or Germany.

Yet while the appointment of local inspectors was expected to ensure auditing was sensitive to local contexts, and with benefits for smallholder farmers, there was other evidence that suggested a Northern bias in audit arrangements. The prioritization of Northern interests was evident in terms of both the content of organic standards and compliance processes. For example, smallholder farmers who were members of the KOPCS expressed frustration at the limited knowledge they were given about the content of standards to which they were compliant, or the requirements of the markets for which their produce was destined. Some farmers lamented the limited knowledge they had about allowable farming practices, with one farmer stating he had avoided experimenting with new farming practices or with new crops for fear of breaching organic standards. In addition, some smallholders expressed frustration at their limited access to information about the processes related to the formation of standards. One farmer described standards and other compliance issues as 'falling from nowhere', as well as 'constantly changing'. This smallholder held a wrinkled black and white photocopied photograph of fruit with blemishes and marks, and of various sizes, and explained this was the 'market information' related to quality and size provided by their export buyer. This farmer lamented:

> How can we know about your markets? How can we know what you want when all we have to go by are such images? I want to go to your country, to learn more about what you *muzungu* (white people) want.

Even with local inspectors, smallholder farmers appear to be left to respond to what they described as an ever-changing organic regimen, which they have little understanding about or input into. Similar frustrations related to standards were also expressed by representatives from NOGAMU and UgoCert, including the limited extent to which local interests were recognized and included in the formation of organic standards, as well as audit and inspection arrangements. Interviews with representatives from UgoCert, for example, revealed the limited success they believed had been achieved in negotiating equivalence between national, regional and international standards. A representative from UgoCert expressed his frustration in this regard: 'We have no bargaining power, we have absolutely no say in international negotiations.'

In sum, the formation of smallholder group certification and local inspection in Uganda has delivered mixed opportunities for participation and outcomes for local actors, including smallholders. This chapter now turns to India, where the formation of a Participatory Guarantee System has delivered some different outcomes for smallholder participation and rural livelihoods.

Participatory Guarantee Systems in India and opportunities for smallholder participation

Organic farming was the traditional method of farming in India until the introduction of the Green Revolution in the 1960s. The development of the concept of organic farming began with Sir Albert Howard's book, *An Agricultural Testament*, which was published in 1940 after his return from India (Howard, 1940; Heckman, 2007). Howard argued that organic farming had been practised in India for many decades – and indeed centuries – before his observations were made. After many years of practising conventional farming, farmers in India are now becoming involved in the movement to give up conventional farming, and switch back to natural or organic farming.

The demand for governance mechanisms for India's organic agriculture sector has grown as the sector has entered the global market. Over the past three decades, the rapid growth in organic production (with India now comprising the largest number of certified organic producers worldwide) has resulted in an increasing emphasis on the certification of organic goods. The outcome of this has brought specific challenges to smallholders.

To begin, and similar to the circumstances described in Uganda, small-scale producers have been confronted with the mainstream organic certification process, which is highly bureaucratic, and which requires producers to follow a code of practice designed by global market agencies. Mainstream organic certification also requires producers to follow a lengthy process of certification – often for just a small quantity of produce – and without economic support (Smith and Lyons, 2011).

This model of organic certification relies on verification via an external certifier who undertakes an audit and evaluation process based on the guidelines built by the certification agencies, and largely without the participation of local-level producers (see also Smith and Lyons, 2011). The expense and paperwork demanded by the organic certifier limits the entry of small-scale producers into organic certification. One participant who was practising organic farming articulated this disadvantage:

> I have a small farm and I will not be able to pay for the certification process. Moreover, I have also observed that there are many certification agencies and if I chose to market to another buyer then I need to do this all over again.

Such circumstances – which limit the entry of smallholder farmers into organic certification – appear to contradict the philosophy and principles of the organic production system; which aims at fair trade and the delivery of fair benefits.

Participatory Guarantee Systems

In order to address some of the limits associated with mainstream organic certification processes, and to incorporate fair trade practices within organic trade, an alternative certification system oriented towards small-scale producers has been devised: the 'Participatory Guarantee System' (PGS). Participatory Guarantee Systems have previously been introduced in other parts of the world, including Brazil, New Zealand and the United States. In 2006, and supported through a number of organic institutions, this system was introduced to India (Khosala, 2006).

The Indian PGS model was developed through the technical cooperation programme initiated by the Indian Ministry of Agriculture and the FAO. The Indian PGS model was developed as an outcome of a process with smallholder farmers, local groups, the regional council and the national coordinating committee, and facilitated by NGOs.

The 'farm family' is the basic unit of the PGS; the farmer, along with his/her family, makes a pledge that they understand and will adhere to organic standards. The 'local group' then refers to a group of farmers who are trained to share information on organic practices, and will take the lead in conducting inspections on neighbouring farms. The 'local groups' in the region coordinate to form a 'regional council', which is in turn facilitated by state agencies or NGOs. The group functioning as the 'National Coordinating Committee' consists of a representative from the Ministry of Agriculture, consumer groups, NGOs and an unlimited number of qualified regional council groups (Khosala, 2006).

The use of PGS as a complementary system of certification to third-party certification has facilitated the entry of a great number of farmers into the certified organic system in India, the outcomes of which have delivered a range of opportunities and challenges for smallholder farmers.

New forms of participation?

The key elements of the Participatory Guarantee System are 'participation, trust, transparency, learning process, horizontality, decentralization, formation of networks, local focus, food security and sovereignty' (IFOAM, 2007).

In terms of participation, and in contrast to the introduction of group smallholder certification in Uganda, the PGS system relies on smallholder farmers themselves conducting processes related to organic inspection and compliance. One smallholder articulated this:

> Actually we do certification among ourselves, which is called participatory guarantee scheme. So we go on (and) mutually check each other's farm. As we are local, we will come to know through someone, or through the labourers, if any chemical was attempted to use without our knowledge.

The PGS model represents a decentralized form of certification, whereby smallholders themselves are able to undertake inspection processes within their local village, and utilizing local knowledge to interpret compliance with organic criteria. These circumstances suggest smallholders have some agency, enabling them to act as moral arbitrators in defining organic farming systems (unlike the group smallholder system described in Uganda – where Southern actors had gained little ground in terms of shaping the content of organic standards).

In addition, the social network in the Indian context is widely believed to be strongly interwoven and transparent in the rural settings, rendering it difficult to hide certain activities (such as chemical use) from other members in the community. In this context, it would lead to embarrassment if any farmer tried to 'cheat', by breaking with organic farming principles. For instance, one farmer stated:

> We know each other very well. Once we have agreed to follow it together (organic farming principles) then we have to keep our promise. We know that some fake produce is available in the market, and we have taken necessary steps to stop this.

In addition to conducting inspection – characteristic of traditional third-party certification – the 'local farmers' group' also provides support and education to smallholder farmers. The PGS model plays a vital role building networks and peer-support systems to assist farmers to gain access to information and problem-solving skills, to ensure their practices comply with organic principles, as well as to prevent instances of cheating. One farmer articulated the benefits of such informal support:

> I had troubles following the organic principles, as I have been using a lot of chemicals in my farm. But my friends from the local group gave me ideas and encouragement to continue to practise organic farming. So my land will be ready for certification by next year.

Informal farmer supports are not unique to PGS; indeed they are also present within organic farming communities certified via third-party and smallholder certification systems (see Lockie et al., 2006). Yet what is distinguished here is the extent to which the PGS identifies certification, inspection and informal support groups all to be part of the quality assurance process, as well as recognizing the central role of smallholders themselves in shaping each of these activities. In this context, and in contrast to the smallholder group certification process in Uganda, the PGS model creates new spaces for smallholders themselves to be actively engaged in defining standards, and ensuring local-level understanding and compliance with these standards.

Discussions about fencing and spray drift demonstrate the extent to which the PGS provides a mechanism for smallholder farmers to provide input into organic standards-setting processes:

> There are some problems with having live fence[4], as this gives shade to the younger coconut trees [and other ground cover crops, resulting in a] reduction in yield. There is another problem with live fence; we are not able to view (our gardens) properly, so there is (the) possibility of theft of the coconuts. So we have told the local farmers' group that we cannot have live fence . . . Moreover, the problem of [spray] drift is less when we have annual crops. If neighbours have short term crops only then would they spray often. But as our neighbours have perennial crops, they do not spray very frequently, so considering this, the certification agency [Organic Farming Association of India] should have some changes in the guidelines as well. We have written to the certification agency about this problem through the local group and we are waiting for their reply.

The structure of the PGS, including 'local farmers' groups' comprising local community members, has enabled smallholders to directly engage with inspectors through an appraisal process. Here, local farmers' groups have provided feedback to the organic certifier about the inappropriateness of live fences in all instances. While the outcome of these negotiations remains outstanding at the time of writing, smallholders themselves expressed some satisfaction that the PGS process provided a new structural arrangement for smallholder farmers' voices to be heard.

Limited inclusion

While the PGS model has succeeded in supporting the increased participation of farmers in organic certification (a stated objective of PGS), at the same time, and similar to concerns raised by smallholders and others in Uganda, access to certification remains constrained. Some farmers, for example, were unable to comply with the organic criteria stipulated by the local farm group. One farmer articulated this:

> I was determined to certify my farm. But I found it very hard to generate organic manures for the farm, as I don't have animals on my farm. Also the practices are labour intensive. Labourers refuse to handle organic manures and sprays. So I had to give up!

In addition, a common assumption underpinning the PGS model is that participation and knowledge are freely available in the village, and that all potential adopters can access it equally (Palaniappan, 2009). Although this is true to some extent – and indeed is articulated among the goals and objectives of PGS – there are various circumstances where such outcomes are not realized. For example, some farmers expected payment for their services, including sharing their learning with other farmers:

> We wanted to learn about vermicomposting. So we contacted a 'local group farmer' in the neighbouring district to demonstrate vermicomposting to us. We had to pay him from our group fund. This was quite different for us as many of us took pride in sharing our experiences with other farmers.

This experience demonstrates the extent to which participation and knowledge sharing vary, and may be shaped by a number of factors, including individual and community expectations, as well as a range of other factors (Palaniappan, 2009). Despite delivering arguably greater levels of participation and inclusion compared to the smallholder certification system in Uganda, there also remained evidence of farmer confusion related to standards and compliance issues, including confusion associated with the existence of multiple certification bodies, including both national and international bodies, with one farmer lamenting:

I know there are a couple of Indian certification bodies and foreign certification bodies. I am told that I need to choose the certification body based on the crop I grow and to whom I want to sell. This is too much for a small farmer like me.

In contrast, in our third case study – the Philippines – the majority of organic farmers are certified via individual third-party accreditation services, with only a few notable exceptions based on a model of group certification. There is also little evidence of participation and inclusion in the certification process, as articulated in the PGS model. Yet despite these factors, the results of a livelihood survey demonstrate conversion to certified organics has delivered clear positive livelihood outcomes for farmers.

Philippines – standards setting, audit requirements and experiences of smallholders

According to official statistics, the area of certified organic agricultural land in the Philippines increased nearly fourfold from 15,795 hectares in 2008 (or 0.13 per cent of total agricultural land) to 52,546 hectares (or 0.45 per cent) in 2009 (Willer and Kilcher, 2011). This placed the Philippines well ahead of other South East Asian countries including Indonesia (0.11 per cent), Malaysia (0.02 per cent), Thailand (0.15 per cent) and Vietnam (0.14 per cent). This recent and dramatic spike in the area of certified organic farmland reflects over a decade of work in the development and promotion of organic standards and farming systems. The first domestic private certification programme was launched by the Organic Certification Center of the Philippines (OCCP) in 2001. In 2003, OCCP began drafting the Philippine National Standards for Organic Agriculture, which were subsequently adopted by the Department of Agriculture's Bureau of Agriculture and Fisheries Product Standards. This was followed, in 2005, by Executive Order 481 which required the Department of Agriculture to establish a National Organic Agriculture Program focused on further development of regulations, certification systems, market promotion, research and so on. While OCCP members are primarily Filipino farmers, food activists, NGOs, businesses and academics, both the original OCCP standard and the National Standards closely mirror model standards developed by IFOAM. No provision is made within these standards for group certification (e.g. Internal Control Systems or Participatory Guarantee Systems). However, a number of peoples' organizations – generally with the support of NGOs – have attained certification to non-domestic standards that do allow group certification. Within the Negrenese *barangays* included in the research reported here, for example, two coffee growers' associations had been formed that utilized Internal Control Systems certified by European organizations.

The majority of households participating in the research on which this chapter draws farmed around one hectare as beneficiaries of the Comprehensive Agrarian Reform Program. Major crops were rice, maize, fruit, vegetables and, to a lesser extent, coffee. Households reported mean annual net incomes only marginally above the rural poverty line. Over half the households were in debt and most nominated a two- to four-month window every year of food insecurity. Many of the households surveyed had been assisted to certify to international organic standards by one of two NGOs: Broad Initiative for Negros Development (BIND) or Negros Island Sustainable Agriculture and Rural Development Foundation (NISARD). A smaller number of households were involved in non-certified production of fair trade bananas.

Livelihood outcomes

One of the key objectives of NISARD and other NGOs in promoting organic certification was to improve farm incomes through access to higher-value export markets. Success on this front among surveyed farmers was limited. Nevertheless, improvement of produce quality had enabled the coffee growers' associations established by NISARD to develop a premium local market for 'Negros Rainforest Coffee' that competed against generally higher-status imported coffees. While more intensive management increased the cost of production and processing, this was more than offset by increased productivity and a doubling in the average price received. Similar results were reiterated by some organic coffee growers in Uganda, who also identified increased yields and improvements in income as an outcome of conversion to organics.

Organic rice farmers, by contrast, displayed similar levels of productivity to conventional rice producers, but achieved net returns 30–50 per cent higher on rice sold due to reduced input costs and premiums offered by the domestic organic market. However, access to price premiums was a less important motivation for organic rice production than was cost reduction and reduced exposure to agrichemicals. Aversion to input use was in fact common amongst both organic and conventional farm households, with many failing to invest adequately in crop nutrition. Yields of staple crops like rice consequently varied enormously among households whether production was organic or nominally conventional, the most important factors explaining this variability being the education and managerial capabilities of household members.

In the context of total net household income, the direct contribution of organic production and certification was, for the majority of households, incremental. Nevertheless, farmers were confident that the benefits of more ecologically friendly practices including organic production, sloping area land technology, etc. were also cumulative and would lead to better long-term livelihood outcomes. Given the low levels of input use among upland farmers, regardless of their organic or conventional status, it is likely that all farmers benefit from more healthy agro-ecologies at a landscape scale.

The activities of NGOs supporting and promoting organic conversion must be seen in a similar context. The assistance of BIND and NISARD for upland farmers in Negros Occidental was not focused solely on certification, but on a variety of interrelated activities including farmer organization, training in organic production methods (including organic fertilizer production), protection and rehabilitation of forested slopes, utilization of locally endemic plant varieties, development of plant genetic resources, biosecurity, market development and alternative livelihood options (a similar list of service provisions was offered as part of the PGS in our Indian case study). Again, the benefits of these activities are likely to be shared among both organic and conventional farmers.

Certainly, there are issues of dependency implicated in the relationships between upland farmers in Negros Occidental, NGOs, buyers and organic certification bodies. Most evident is the dependency of upland farmers on indigenous NGOs operating at provincial and national scales. NISARD, BIND, OCCP and many other relevant NGOs draw their primary membership from local elites; well-educated and comparatively wealthy landowners and other business people with the skills and political connections to solicit support for their activities from government and international agencies. These NGOs also employ a significant professional staff. Thus, while the Philippine National Standards for Organic Agriculture make little provision for the particular needs of resource-poor small farmers, these NGOs are more than capable of identifying international standards and certification systems that do meet these needs for export purposes. They have also proven themselves adroit in establishing relationships of trust with buyers, particularly within domestic markets, to supply non-certified and/or 'self-certified' organic produce. Certification could not, therefore, be described by itself as a major constraint on the livelihood options or autonomy of small farmers.

Conclusion: governing organic agriculture in the Global South and livelihood outcomes

Organic production has expanded significantly in the Global South since the year 2000. Uganda, India and the Philippines, the case studies discussed in this chapter, have all experienced significant organic sector growth. The majority of Southern organic produce, including that grown in our selected case study countries, is sold in Northern markets. The expansion of South–North international organic commodity chains is mediated by organic governance arrangements that provide verification between producers, consumers and other intermediary actors. The traditional model of organic governance – including individual third-party accreditation – has been increasingly critiqued for its social and ecological insensitivities regarding the specific needs of smallholder Southern farmers. In response to such concerns, a number of alternative smallholder-specific and participatory certification schemes have emerged including small-holder group certification and the Participatory Guarantee System (PGS). In Uganda, smallholder-specific group certification has been vital in facilitating the expansion of the organic industry. Similarly, in India, the PGS model has facilitated the increased entry of farmers into organic certification. In contrast, the majority of certified organic farmers in the Philippines rely on individual third-party accreditation, with just a few cases of group certification.

The expansion of both international organic trading and localized models of organic governance are compatible with broader agricultural development discourses, including export–led, participatory and sustainable development narratives. But to what extent have smallholder farmers and other local-level actors actually participated in shaping organic governance mechanisms? And what are the social, economic and ecological outcomes associated with the uptake of organic agriculture in the Global South? Our results paint a mixed picture in relation to each of these questions.

To begin, the introduction of smallholder group certification and PGS has clearly provided new opportunities for Southern smallholders to enter into organic production and international organic commodity trading. Individual third–party accreditation previously rendered organic certification prohibitive for most smallholders. Yet while smallholder certification has created new opportunities for smallholders to participate in organic trading, our results point to the limited extent to which local-level actors have actually been engaged in processes that might shape governance institutions. The results from our Ugandan case study, for example, show that Northern stakeholders largely defined the smallholder certification scheme – including organic standard-setting and inspection processes. Similarly, in both India and the Philippines, Southern elites (including representatives from local farmer groups and NGOs) were central in both defining the content of and interpreting organic standards. However, Ugandan local actors were successful in advocating for local inspectors; an outcome that has delivered clear economic benefits to export companies. Meanwhile, the Indian PGS model has provided a structure for local farmer groups – and via engagement with smallholder farmers – to provide input into standards-setting processes. The PGS has also widened the scope of organic governance, including training and farmer support alongside traditional inspection roles. Meanwhile in the Philippines, there were few examples of group certification, with most organic farmers certified via individual third-party accreditation. While such arrangements appeared to create new forms of dependency (both South–South and North–South), the livelihood data demonstrated a range of social, economic and ecological benefits that flowed from participation in certified organic markets.

While 'participation', 'empowerment' and 'engagement' are frequently touted in a positive light in agricultural development discourse, our results demonstrate the problematic nature of utilizing such terms to evaluate agricultural development strategies and their outcomes. First, the uncritical use of such terminology may mask the power relationships that prohibit effective participation and democratic engagement; including the power relationships that limited smallholders and other Southern actors in shaping emerging organic governance arrangements. Second, our results demonstrate a range of positive socio-economic and ecological outcomes associated with the uptake of organic agriculture, including in circumstances where there was little effective participation among smallholders and other local-level actors in organic governance arrangements. While there were various constraints upon smallholders and other local-level actors' participation in shaping the form and outcomes of organic agricultural development, entry into organic agriculture has delivered a range of positive social, economic and environmental outcomes.

The results presented in this chapter point to some of the challenges related to organic governance. These include the task of establishing broadly inclusive models of organic governance that are able to place the lived realities of Southern smallholder farmers and other local-level actors – including their hopes and aspirations – at the forefront of organic governance debates. This repositioning will be vital to ensure that the Global South's organic sector expands in ways that are compatible with broader goals of ensuring a socially and environmentally just agri-food system.

Notes

1   It is likely that the non-certified organic sector in the Global South is much larger than the certified organic sector. However, there are currently limited data related to the non-certified sector.
2   Local box schemes are one of a number of alternative food networks that provide direct connections between producers and consumers. The Kenya Institute of Organic Farming introduced the local box scheme that was the focus of this study in 2007 to connect smallholder farmers and consumers living in Nairobi, the capital (Freidberg and Goldstein, 2010). Freidberg and Goldstein (2010) identify various challenges that acted to constrain the long-term viability of this local food initiative, including both ideologies of conventional development, as well as institutions, policies and activities that have prioritized export agriculture.
3   The data presented in this chapter by Palaniappan is drawn from her PhD study entitled 'Learning to practice transitional agriculture: an action research thesis' (2009), University of Queensland.
4   Border trees like Neem and Pongamia are grown on the borders of the farm as a 'live fence'. This is different from a wire or metal fence. The live fence was recommended by third-party organic certification bodies to avoid pesticide drift from neighbouring farms, by providing a wind brake between properties. In addition, the live fence is also able to provide biomass for green leaf manuring and composting, as well as serving as perches for birds that help in management of insect pests.

References

Bacon, C. (2008) 'Confronting the Coffee Crisis: Can Fair Trade, Organic and Speciality Coffees Reduce the Vulnerability of Small-Scale Farmers in Northern Nicaragua?', in C. Bacon, V. Ernesto Mendez, S. Gliessman, D. Goodman and J. Cox (eds), *Confronting the Coffee Crisis: Fair Trade, Sustainable Livelihoods and Ecosystems in Mexico and Central America*, Cambridge, MA: MIT Press, pp. 155–178

Barrett, H., Browne, A., Harris, P. and Cadoret, K. (2002) 'Smallholder Farmers and Organic Certification: Accessing the EU Market from the Developing World', *Biological Agriculture and Horticulture,* 19, 2, pp. 183–199

Campbell, H. and Le Heron, R. (2007) 'Supermarkets, Producers and Audit Technologies: The Constitutive Micro-politics of Food, Legitimacy and Governance', in D. Burch and G. Lawrence (eds), *Supermarkets and Agri-food Supply Chains:*

Transformations in the Production and Consumption of Foods, Cheltenham: Edward Elgar, pp. 131–153

Department for International Development (DFID) (2001) *Sustainable Livelihoods Guidance Sheets,* London: Department for International Development

Dolan, C. (2010) 'Virtual Moralities: The Mainstreaming of Fairtrade in Kenyan Tea Fields', *Geoforum,* 41, 1, pp. 33–43

Food and Agriculture Organization (FAO) (2007) *International Conference on Organic Agriculture and Food Security.* Rome: Food and Agriculture Organization of the United Nations.

Freidberg, S. and Goldstein, L. (2010) 'Alternative Food in the Global South: Reflections on a Direct Marketing Initiative in Kenya', *Journal of Rural Studies,* 27, 1, pp. 24–34

Friedmann, H. (2004) 'Feeding the Empire: Pathologies of Globalised Agriculture', in L. Panitch and C. Leys (eds), *Socialist Register: The Empire Reloaded,* London, Merlin, pp. 124–143

Friedmann, H. and McNair, A. (2008) '"Whose Rules Rule?" Contested Projects to Certify "Local Food Products for Distant Consumers"', *Journal of Agrarian Change,* 8 (2 and 3), pp. 408–434

Heckman, J. (2007) 'A History of Organic Farming: Transitions from Sir Albert Howard's War in the Soil to USDA National Organic Program', *Renewable Agriculture and Food Systems,* 21, 3, pp. 143–150

Howard, A. (1940) *An Agricultural Testament.* London: Oxford University Press

IAASTD (2008) 'International Assessment of Agricultural Knowledge, Science and Technology for Development'. Available on-line at: www.agassessment.org (accessed 5 May 2010)

International Federation for Organic Agriculture Movements (IFOAM) (2007) 'Participatory Guarantee Systems Shared Vision, Shared Ideals'. Available on-line at: www.ifoam.org/about_ifoam/standards/pgs/pdfs/IFOAM_PGS_WEB.pdf (accessed 10 March 2011)

Jaffee, D. (2007) *Brewing Justice. Fair Trade Coffee, Sustainability and Survival.* Berkeley: University of California Press

Katto-Andrighetto, J. (2011) 'Government Recognition of Participatory Guarantee Systems in 2010', in H. Willer and L. Kilcher (eds), *The World of Organic Agriculture – Statistics and Emerging Trends 2011,* Bonn and Frick: IFOAM and FiBL, Frick, pp. 82–83

Khosala, R. (2006) 'A Participatory Organic Guarantee System for India'. Report Available on-line at: www.ifoam.org (accessed 10 April 2010)

Kung Wai, O. (2011) 'Organic Asia 2010', in H. Willer and L. Kilcher (eds), *The World of Organic Agriculture – Statistics and Emerging Trends 2011,* Bonn and Frick: IFOAM and FiBL, Frick, pp. 122–127

Langan, M. (2011) 'Uganda's Flower Farms and Private Sector Development', *Development and Change,* 42, 5, pp. 1207–1239

Lockie, S., Lyons, K., Lawrence, G. and Halpin, D. (2006) *Going Organic: Mobilising Networks for Environmentally Responsible Food Production,* Wallingford, UK: CABI Publishing

Lockie, S., Tennent, R., Benares, C. and Carpenter, D. (2010) 'Subsistence, Food Security and Market Production in the Uplands of Negros Occidental, the Philippines'. Unpublished Report. Australian National University, Canberra.

Lyon, S., Bezaury, J. and Mutersbaugh, T. (2010) 'Gender Equity in Fairtrade-organic

Coffee Producer Organisations: Cases from Mesoamerica', *Geoforum*, 41, 1, pp. 93–103

Lyons, K. (2012) 'New Opportunities to Restructure and/or Re-Imagine Export-Led Markets? Organic Farming in Ghana and Uganda', *Journal of Agrarian Change*

McMichael, P. (2004) *Development and Social Change: A Global Perspective*, (4th edn). Thousand Oaks, CA: Pine Forge Press

Mutersbaugh, T. (2004) 'Serve and Certify: Paradoxes of Service Work in Organic-Coffee Certification', *Environment and Planning D: Society and Space* 22, pp. 533–552

Mutersbaugh, T. and Klooster, D. (2010) 'Environmental Certification: Standardization for Diversity', in S. Lockie, D. Carpenter (eds), *Agriculture, Biodiversity and Markets: Livelihoods and Agroecology in Comparative Perspective,* London: Earthscan, pp. 155–174

Organic World Net Philippines: Department of Agriculture to Support Organic Agriculture (2010) Available on-line at: www.organic-world.net/35.html?&tx_tt news[tt_news]=430&cHash=dbfa0a2eed (accessed 15 March 2011)

Namuwoza, C. and Tushemerirwe, H. (2011) 'Uganda: Country Report', in H. Willer and L. Kilcher (eds), *The World of Organic Agriculture – Statistics and Emerging Trends 2011*, Bonn and Frick: IFOAM and FiBL, pp. 117–120

Palaniappan, G. (2009) "Learning to Practice Transitional Agriculture", PhD thesis, University of Queensland, Australia

Parrott, N. and Marsden, T. (2002) *The Real Green Revolution. Organic and Agroecological Farming in the South.* London: Greenpeace Environmental Trust

Smith, K. and Lyons, K. (2011) 'Negotiating Organic, Fair and Ethical Trade: Lessons from Smallholders in Uganda and Kenya', in C. Campbell, C. Rosin and P. Stock (eds), *Food Systems Failure,* London: Earthscan, pp. 180–199

Thamaga-Chitja, J. and Hendriks, S. (2008) 'Emerging Issues in Smallholder Organic Production and Marketing in South Africa', *Development Southern Africa,* 25, 3, pp. 317–326

UNCTAD-UNEP (2008) *Organic Agriculture and Food Security in Africa,* United Nations Conference on Trade and Development, United Nations Environment Program, New York and Geneva

van Elzakker, B. and Rieks, G. (2003) 'Smallholder Group Certification. Compilation of Results'. Proceedings of Three Workshops (February 2001, February 2002, February 2003), IFOAM – Smallholder Group Certification, Germany

Walaga, C. and Houser, M. (2005) 'Achieving Household Food Security Through Organic Agriculture? Lessons from Uganda', *Journal Für Entwicklungspolitic,* 11, 3, pp. 65–84

Willer, H. and Kilcher, L. (eds) (2011) *The World of Organic Agriculture – Statistics and Emerging Trends 2011*, Bonn and Frick: IFOAM and FiBL

7 The possibilities for inclusion of smallholder farmers in organic market chain development

Paul Rye Kledal, Frank Eyhorn, Bo van Elzakker and Elsio Antonio Pereira de Figueiredo

Introduction

Global demand for organic products has remained robust, and is estimated to have reached US\$ 54.9 billion in 2009 tripling the value from US\$ 18 billion in 2000. Most of the sales take place in the USA and Europe (97 per cent), with a growing commodity import from the developing countries (Willer and Kilcher, 2011). However, the rapid rise of both supermarkets and an urban upper-middle-class consumer segment among emerging economies and developing countries (Reardon and Berdequé, 2003) has created a recent important expansion of the domestic market in these countries, leading to progressive transformations and a growing structural complexity of their respective organic sectors.

Both FAO and IFAD (Scialabba, 2007; IFAD, 2005) see a promising opportunity for smallholders in developing countries to increase their incomes and access a better livelihood in benefiting from the booming of the organic demand. Many research results from countries like India, Tunisia, Turkey, Cuba or Sub-Saharan African countries confirm these optimistic views. They show that organic farming effectively has the potential to provide smallholders with access to attractive markets with higher profitability, while creating new partnerships within the whole value chain and strengthening self-confidence and autonomy of the farmers (Crucefix, 1998; Shah et al., 2005; Kilcher, 2007; Bolwig et al., 2009; Källander and Rundgren, 2008).

Despite these potential prospects for smallholders of moving into the organic market segment, it does not automatically lead into a 'safe haven' free from classical market rivalry and exclusion. Gomez-Tavar et al. (2005) and Gonzalez and Nigh (2005) show how in Mexico the requirements to be certified organic and the highly competitive market for organic coffee tend to reproduce social inequalities between smallholders and larger commercial market-oriented producers, triggering exclusion of the former. Similarly, Blanc (2009) shows how the Brazilian domestic market for organic vegetables is highly competitive and exclusive and does not protect smallholders from potential hold-up[1] situations exerted by powerful downstream partners. From a broader perspective, research underlines that smallholders from emerging countries – often economically marginalized and with low educational backgrounds – are facing many challenges to enter the organic sector and benefit from it. Problems such as decreasing incomes during the conversion period, high costs for certification and competences in managing the certification requirements are serious constraints, particularly when no specific subsidies exist for organic production (Egelyng, 2009). Structural barriers to access credit, difficulties in creating reliable market linkage and lack of knowledge of organizational management are addressed as highly problematic issues too (Barrett et al., 2001; Nordlund and Egelyng, 2008; Blanc, 2009), but could also be regarded as common development issues and not particularly related to organic production.

This chapter aims at further discussing these issues and presenting possible approaches to overcome the broad range and type of existing market and inclusion barriers. For specific aspects of power inequalities and certification schemes, see Chapter 6. The chapter itself will be structured around:

- A conceptual framework of the four agro-food systems prevalent in developing countries and their type of economic market chain organization leading to inclusion as well as exclusion of smallholders.
- A theoretical and analytic framework based on New Institutional Economics explaining the type of costs and barriers smallholders endure when trying to enter into more commercial market chains and food systems.
- A catalogue of possible approaches to overcome smallholder market inclusion barriers.
- A concluding part summarizing the major issues put forward in this chapter.

The four agro-food systems prevalent in developing countries

As illustrated in Figure 7.1, four different typologies of agro-food systems, closely related to the development process taking place among developing countries, can be categorized. The global growth and trade of organic food commodities are closely related to these four agro-food systems.[2]

The first is classified as a *traditional* agro-food system, characterized by a dominance of short supply chains, connected with many middlemen and limited market structure. Smallholders, presented by the small dots in the figure, deliver to a nearby town and a traditional wholesale market where shopkeepers of the classical family 'mom and pop shop' attend to buy the farmers' produce. Most of the world's poor smallholders live within this food system. There is not much to improve; the prospects of development in terms of transfer of knowledge, new technology and rising income is related to the potential inclusion of poor smallholders into the other three (expanding modern) agro-food systems.

The second and third systems are the formalized procurement systems pictured on the right side of Figure 7.1, consisting of a *domestic-oriented* and an *export-oriented* agro-food system.

The domestic system is still characterized by traditional market actors, but with a more complex set of rules and regulations applied to marketplaces and a

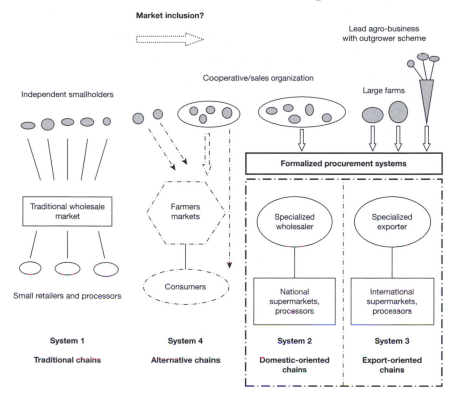

Figure 7.1 Organic value chains and their links to the four agro-food systems
prevalent in developing countries

higher degree of market infrastructure. Within the domestic-oriented food
system organized supply chains operated by national and/or international
supermarkets are capturing a growing share of this market, with urbanization as
its underlying driver. However, for smallholders to enter this food system 'the
critical mass of supply' is essential, because modern supermarkets demand stable
and secure supplies for their outlets. To be able to deliver the required quantity
and quality during a significant part of the year smallholders therefore have to
engage themselves in some sort of a market association or a cooperative, as
illustrated in Figure 7.1 with the individual smallholders placed in an ellipse.

The export-oriented agro-food system can be characterized as a highly
industrialized agro-food system found throughout the developed world. It holds
strong perceptions of food safety, a high degree of coordination and a large and
consolidated processing sector with organized retailers. The *export* market in the
emerging countries is oriented towards this agro-food system driven by demand
mainly from the developed world. In general, the demand system prevalent in
the export-oriented system only trade directly with a few large and dedicated
lead farmers or market associations, as illustrated in Figure 7.1 with the triangle

in the upper right. However, as a supply buffer, especially within Fresh Fruit and Vegetables (FFV), these large suppliers have their own packaging, storing and cooling houses and contract outgrowers, sometimes as core suppliers, sometimes as standby for their sales. This type of supply-chain organization with an outgrower scheme is at times utilized by development agencies or corresponding NGOs as a way to include smallholders in modern export and value chains, and hence upgrade the farmers in terms of income as well as transfer of knowledge and new technology.

The fourth agro-food system is named an *alternative* food system, where farmers, various types of intermediaries and consumers are able to construct semi-closed circuits of exchange. These semi-closed circuits are often based on values stressing transactional processes of trust, community, social and environmental welfare, as against capitalist transaction outcomes of exchange such as competition, exclusion, price decline and concentration of production. Market exchange can be global as well as local. The latter is often organized as fairs or direct delivery through box schemes and the farmers can be either operating individually or through some kind of market organization, as illustrated in Figure 7.1.

The growth of organic production in developing countries is first and foremost linked to the export-oriented agro-food system. However, the rapid rise of both supermarkets and an urban upper-middle-class consumer segment in emerging countries has opened up sales for organic food suppliers to expand through the formalized and domestically oriented food systems found in countries such as Brazil, China and Egypt. Consumer values are found to be similar to the ones documented in the developed world, with individual concerns towards health as well as altruistic values regarding the overall environment (Siriex et al., 2011).

The general expansion of organic sales through the domestic formalized procurement system has so far been driven by large global supermarket chains like Carrefour (home base: France), smaller global chains like Casino and Metro (home base: France and Germany) and up-coming regional players like Grupo Pão de Açúcar (Brazil), Hyper One (Egypt), Uchumi and Nakumatt (Kenya), Lianhua, Hualian (China) and Shanghai Nong Gong Shang (Shanghai-based) (Kledal et al., 2007, 2008, 2009), and in South Africa chains like Woolworths, Shoprite, Pick & Pay. Likewise, organic sales are so far all found to be concentrated in high-end supermarkets placed in upper-middle-class residential areas in the big metropolitan centres.

The *alternative organic market*, is also expanding worldwide often making its existence on two types of consumer base. One is locally based and the value chains are short resembling to some degree the market structure found in the traditional food system; and the other is targeting the urban consumers alternating between the outlets of the second food system and/or the alternative food system. In Brazil the alternative food system is estimated to cover almost 50 per cent of the domestic organic food market via popular fairs, box schemes or direct delivery systems (Willer and Yussefi, 2006), and is deeply connected to the historically strong agro-ecology movement of the country (see Chapter 10).

The horizontal dynamics between the four agro-food systems

The perspective of the four food systems should not be viewed as only operating on a classical vertical market chain level, but also on a horizontal level. The various food systems and value chains are all socially and economically embedded into either rural, regional or global settings as illustrated in Figure 7.2. When the focus is on smallholder inclusion and market commercialization the challenges therefore differ immensely if the value chains are operating in 1) an export-oriented food system trying to connect poor smallholders embedded in a remote rural setting to a global market; 2) commercial smallholders operating in the domestic urban–market–oriented food system; or 3) semi–commercial smallholders in a rural setting upgrading their produce for expansion to new alternative or existing rural markets.

The conceptual framework of the four agro-food systems presented here is therefore first and foremost to give an overview and an understanding of the complexity of the problems smallholders face when trying to be included in more commercialized food and market systems. Examples of this will be presented in more detail in the following paragraphs. However, the structure and framework of the four agro-food systems and their type of economic market organization is not to be understood as a static conception of how food markets operate. In many cases smallholder farmers, individually or as a group, alternate between the various agro-food systems. For example, products sold for the export market are normally subject to very high quality requirements as well as a specific quantity. Excess amounts or products divergent in quality requirements are then often sold to local domestic supermarkets, but normally at a lower price.

In the case in Box 7.1, the Mobiom famer cooperative in Mali is an example of smallholders producing not only various products but also marketing these products to the various food systems operating under the alternative, domestic- and export-oriented markets.

In the following section the concept and theory of transaction cost from New Institutional Economics (NIE) will lay the ground for explaining the type of costs and barriers smallholders endure when trying to enter into more

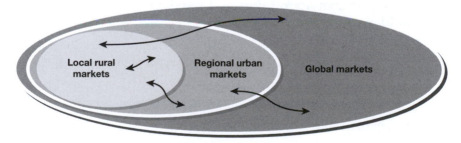

Figure 7.2 The interrelation between smallholders' social and geographical placement and their type of market inclusion

Source: Elzakker and Eyhorn, 2010 (modified).

Box 7.1 Product and market diversification – the Mobiom cooperative in Mali

Mobiom, a federation of organic producer cooperatives in Mali, organizes some 6,000 smallholders for organic production and marketing. With support from the Swiss NGO Helvetas, the organization started with a focus on cotton for organic and fair-trade export markets, but gradually diversified both crops and market channels. The surplus of cereals, pulses and vegetables grown in rotation with cotton is sold in the local market (i.e. in the traditional food system). Some first attempts to sell organic cereals and vegetables to middle-class consumers in the capital Bamako showed encouraging results. Sesame and peanuts are sold under the organic label to exporters for processing into edible oils. Mobiom also organizes the marketing of shea nuts and shea butter from trees grown in certified organic fields. The majority of the shea nuts are exported in bulk to a processor in Burkina Faso, while the remaining are locally transformed into butter in their own processing plant. This high-quality butter is then sold to local as well as to urban consumers. First-quality mangoes are exported as certified organic and fair-trade fresh mangoes, while inferior qualities are either consumed locally or are processed into dried mangoes, which then again are either exported or sold locally. Even in the case of cotton – a typical export commodity – a small part is processed locally in artisanal workshops. The textiles are then sold in urban markets, or are exported.

---

commercial market chains and food systems, and hence draw attention to the huge task many rural development initiatives face as well as to the approaches applied to create sustainable solutions.

The theory of transaction cost and barriers of smallholder market inclusion

To understand the context in which smallholder farms operate Pingali and Rosegrant (1995) classify smallholders into three types of production systems:

- subsistence farming
- semi-commercial
- commercial farming conditions.

Increasing commercialization shifts farm households away from traditional self-sufficiency goals towards profit- and income-oriented decision making. Hence farm output is accordingly more responsive to market needs with increasing commercialization. As the level of commercialization increases, mixed farming systems give way to specialized production units for the production of high-value crops and livestock products (Pingali et al., 2007). The challenges of smallholder inclusion will therefore differ in relation to their type of production system as well as to what agro-food system inclusion is targeted. NIE as an economic theory not only explains the different type of challenges smallholders will endure when entering the various agro-food systems, but also the different modes of firm organizations and institutions ('rules of the game') the actors along a value chain will engage in to reduce those challenges.

NIE uses the 'transaction' as the unit of analysis. From an NIE point of view, exchange itself is costly, meaning that, conversely to the proposition of orthodox economics, the behaviour of market actors cannot be explained and predicted only in considering trade-offs between prices and production costs. NIE thus claims that taking into account the cost that actors face when trying to coordinate their exchange on the market is essential to under-stand individual and collective behaviours in this arena. These costs, called transaction costs, include the costs to obtain and process market information (information costs), to negotiate contracts with others (bargaining costs), to make sure the other party sticks to the terms of the contract (monitoring costs) and to take appropriate action if this turns out not to be the case (enforcing costs). Hobbs (1997) classified these costs so information costs typically arise *ex ante* of an exchange, bargaining costs are the costs of physically carrying out the transaction while monitoring and enforcement costs occur *ex post* of a transaction.

Williamson (2000) and Masten (2000) define for their part five determinants of transaction costs: (1) *frequency* that refers to how often a transaction takes place, hence helping to build trust and lower monitoring costs; (2) *asset specificities*, e.g. farmers planting coffee or orange trees will have to wait a number of years before selling their fruits and harvest over several years before investments are returned on their (crop-specific) asset. They therefore have to try and minimize their risks *ex post* by making longer-term contracts raising transaction costs *ex ante* to their specific investment; (3) *uncertainty*, the higher the uncertainty about exchange conditions (as in the coffee or orange case above) the more complex contracts will be installed or tensions between market actors can occur all raising transaction costs; (4) *limited* or *bounded rationality;* contrary to neoclassical economics, NIE claims there are limits to how much market actors can know or foresee, hence raising negotiation and monitoring costs in an effort to minimize uncertainty; and lastly (5) *opportunistic behaviour* or *moral hazards* is potentially inherent in humans when it comes to dividing scarce resources, hence making it necessary to install certain safeguarding institutions of control, monitoring and enforcement, all leading to higher transaction costs. A classical opportunistic behaviour evident in agriculture is the so–called 'hold-up' problem. A farmer ready to sell his/her fresh oranges is suddenly confronted with a trading partner not willing to pay the full amount as agreed upon. The farmer has to accept, because s/he will not have time to find a new trading partner before his/her fruit is rotten – a situation known to the buyer. Hence, the opportunistic buyer expropriates part of the return on the farmer's asset-specific investment *ex post*.

According to the seminal work of Coase (1937), it is precisely to economize on these determinants of transaction costs that institutions ('rules of the game') and organizations, internal as well as external to firms, are created. Farm associations or cooperatives are examples of different business organizations where farmers themselves try to minimize various types of transaction costs in terms of securing both the 'critical mass of supply' as well as safeguard themselves against potential 'hold ups'. It can also be seen as a vertical integration downstream of the chain where farmers themselves take over the trading of their product, because the transaction costs of uncertainty, opportunistic behaviour and the asset specificities are too high.

Likewise, a national law on organic certification, control and enforcement and providing support for marketing illustrates how the state as an *external* institution can build trust among producers and consumers and lower their *internal* transaction costs on control and monitoring and hence lower the market price for both parties.

However, setting up organizations and institutions creates internally a set of specific coordination costs. These are related to tasks such as defining and agreeing upon a marketing strategy, establishing collective production and delivery planning, as well as specific governance and property rights arrangements within for example a farm association. Transaction and coordination costs are interrelated and, from an organizational theory perspective, phenomena such as the emergence of an organization, its transformation and strategic moves, as well as its closure, are critically related to the dynamic balance between these two kinds of costs parallel to other basic economic data (demand/offer, price competition, etc.). According to this theoretical perspective, farmers will never build or enter a specific organization if the sum of the costs of transaction and coordination is higher than the expected profits. Conversely, existing organizations will dissolve if these costs become higher than the profits farmers could expect in another market setting (Blanc and Kledal, 2012).

Asset specificities themselves are very important for understanding farmers' willingness to take risk and indulge in a market commercialization. They can be classified into:

• location specificities
• crop specificities
• household specificities.

The location of a farm matters in terms of going commercial. If there is good infrastructure and efficient communication facilities, search and information costs will tend to be relatively lower than in areas with bad infrastructure and inefficient facilities. Likewise a supporting resource environment facilitating high-value crops and commercial markets for inputs and sales will minimize risks associated with a switch to such crops. Similarly the condition of the soil, possibilities for irrigation and fertilization have to be in place for the farmer to calculate the risk.

Asset specificities related to perennials have been touched upon, but seasonal high-value crops like certain fresh fruit and vegetables are often perishable and can therefore be associated with high transaction costs stemming from specific investments in refrigerated transportation, packaging, wastage, quality monitoring and storage costs.

There are a number of household-specific variables that are not so much asset specific per se (Pingali et al., 2007), but significantly impact on them, such as:

• social networks and organization
• age, gender and education
• intra-household interaction.

These variables all influence the costs of information seeking, negotiating, monitoring and enforcement, if one looks at smallholders in a process of moving from subsistence farming to commercialization.

The prevalence of social networks, farmers' or community organizations experienced in horizontal coordination may help to reduce such transaction costs substantially especially if farmers, NGOs or private export companies want to secure the 'critical mass of supply'. The creation of such intermediaries between farmers, traders and/or supermarkets can make the smallholders attractive as a group to trade with and efficient horizontal coordination can secure both the quality as well as consistency on the demand side.

The challenges along the agro-food value chain

As illustrated in the previous section, the range of transaction costs smallholders can endure, *ex ante* and *ex post* of a transaction are many and diverse. In the organic sector where individual supplies can be small and market fluctuations at times greater compared to the conventional sector, the classical paradox of so-called 'market failures', where transaction costs are higher than expected revenues, is a central challenge for starting business in developing countries. An agro-chain partnership among relevant stakeholders as well as a supporting resource environment via state institutions, advisory services or NGOs are therefore essential to reduce and target the many types of transaction costs smallholders endure when trying to build up new organic markets and get them successfully included into commercial markets and relevant food systems.

Table 7.1 has been drawn up to structure and better understand the many constraints existing in an agro-food chain (Eenhoorn and Becx, 2009) and actual transaction costs (Kledal, 2009). It consists vertically of the Supply system, the Intermediaries, the Demand system and the Supporting Resource Environment. Horizontally it distinguishes between 'Common development constraints' and the ones 'Specific organic development constraints'. The Supply system consists

Table 7.1 Common as well as specific development constraints faced by conventional and organic smallholders in emerging countries, when aiming for inclusion into modern agro-food systems

	Common development constraints	Specific organic development constraints
Supply system (smallholder farmers)	Low education levels (in agriculture as well as literacy and counting) Distance from markets Low output – fragmented supply Low risk taking Producer uncertainty on expected price premiums Limited organization/disloyalty to marketing organization	Limited and expensive organic inputs Cost of training Use of ICS for farm improvement Limited adaptation of organic techniques, leading to low production Coexistence with conventional agriculture Cost of certification (when applicable)
Intermediaries (farm co-ops, middlemen, wholesalers, large individual farms/packers, exporters)	Inadequate management capacity Lack of trade finance Lack of quality management, lack of willingness to pay more for quality Limited loyalty up and downstream Focus on own risks, leading to high margins High risk on capital borrowing Conformity assessment costs Staff leaving	Sufficient economy of scale Organic is often a small segment among conventional businesses Cost to establish and maintain internal control systems High search/information/control costs *ex ante/ ex post* on production and exchange when supplying niche markets Comingling with conventional
Demand system (restaurants, processors, specialized exporters, shops, supermarkets)	International competition Small/large order Ever-changing standards and quality requirements Changing consumer markets Asymmetric bargain/trading power	Need for transparency/integrity, leading to vertical integration versus speculation Due to small segments, export markets can be volatile during start-up Different standards in different regions of the world Lack of commitment to producers
Supporting resource environment (government agencies, NGOs, advisory services, banks/credit institutions)	Weak infrastructure (roads, communication) High bureaucracy for firm registration and receiving credit Ownership of land	Relevant research Lack of institutional 'set-up' forwarding information and data on export markets Lack of suitable support policies Organic trading specificities

Source: Adapted from Kledal, 2009.

of the farmers in an agro-chain. The Intermediaries can be farmer cooperatives, processors and wholesalers, and they are an important node in a chain since they link supply and demand. The Demand system consists of the various types of outlets procuring the agro-food products, and are an important node in terms of 'translating' and specifying to the nodes upstream the various attributes that consumers desire of a product. The Supporting Resource Environment would be government agencies, NGOs, advisory services, financial agencies: all important institutions in supporting an effective inclusion of smallholders into commercial agro-food systems.

Focusing specifically on the organic constraints, in the *supply system* it could be inputs such as organic fertilizers or biopesticides, which are often very expensive or difficult to get. For large industrialized export-oriented farms it is a constraint if access to manure as well as controlling outbreaks of pest attacks is difficult. It is seldom a problem for smallholders who base their production on good agro-ecological practices, and with limited crop export. Depending on the contract scheme, producers can have uncertainty both about obtaining organic premiums as well as safeguarding them against fluctuating farm gate prices. The level of premiums and fluctuations is again related to the level of risk the farmers will take and therefore also corresponds to their willingness to make asset-specific investments or commitments and hence adapt to new organic techniques.

Within the *intermediary system* the level of managerial and technical competences (bounded rationality) often limit the benefits from exporting while possibly increasing its costs. Likewise organic exports are often found to be a smaller part of conventional exporters' business, which in some cases sees the organic section getting less priority in terms of resources and attention. On the other hand, the experience inherent in existing conventional companies when introducing organic products means that these companies have a head start compared to the 'pure' organic companies with no experience in export and trading.

Many companies are often undercapitalized yet organic production and processing require specialized know-how and in some cases technology and storage which call for specific investment. However, the financial cost of trading is one of the major constraints among the intermediaries, for example an intermediary can have an annual budget for operational costs of US$40–50,000 spread over the year, whereas the trade finance required may be US$200,000, paid at once. Smallholders may need payment upon delivery, yet the intermediary will only receive payment 45 days later after delivery requiring solid credit schemes for trading. Similarly, volumes are often low in the organic industry, risks not clear and returns are not always enough to encourage firms to make the necessary investments to engage in the organic trade.

In the *demand system* the organic exporters sometimes get small order quantities when they are targeting retail consumer markets that make it difficult to fill a whole container, thus making single product shipments uneconomical. However, many of the small exporting companies would not be able to satisfy a stable frequency of large orders to the retailing industry by themselves. They will have to manage a range of products to allow for regular economical shipping.

Ever-changing quality, food safety and certification standards can be a barrier to entry that requires capacity building for exporters to have dedicated staff that are monitoring changes downstream and translating these into actions upstream.

In the *supporting resource environment* the absence of a national law as well as serious political attention on the macro level creates a lack in funding for wider sector analysis, regular data collection, support for trade fairs, product innovation, etc. Similarly, lack of dedicated relevant research in organic production is a large constraint as well as the challenges of a weak physical and communicative infrastructure.

In the following section possible approaches to overcome the various constraints and transaction costs for organic market inclusion will be presented.

Possible approaches to overcome market and inclusion barriers

The long list of potential challenges and constraints outlined above indicates that successful inclusion of smallholders in market chains is not an easy task. Producing and selling agricultural products, whether with or without smallholder involvement, in the end remains a business decision. Some farmers or farmer groups succeed in dealing with the multiple challenges that a business has to face, and manage to compete in the market, while others don't. Whether an initiative to link smallholders to markets succeeds or not is mainly a question of sustainable competitive advantage, of sufficient scale of operation and of appropriate management capacity to coordinate transactions along a value chain and reduce their costs.

As each country, each product and each market segment has its own particularities it is difficult to identify an approach that works for all possible situations. Nevertheless, some generic guidelines can be deduced from a range of practical examples of organic value chains that involve smallholders. These practical guidelines are outlined below in relation to the challenges, constraints and transaction costs identified in the previous section.[3]

Designing flexible supply systems

Certain constraints related to the organic supply systems diminish when good organic farming practices are applied in diversified farming systems. It is therefore important that the production system is built on methods like crop rotation, mixed cropping and active soil fertility management. An 'organic by default' approach that simply stops the application of inputs prohibited by organic standards will most likely result in secondary problems such as low soil fertility, low yields, pest and disease problems, quality issues, etc. Similarly, a farming system that has an exclusive focus on one product is more prone to adverse production conditions and to market volatility than a diversified system. In addition, it will be more difficult to cover costs for extension and certification if they need to be borne by one crop alone. Smallholders should therefore be encouraged to transform their farms into diverse and productive entities that are able to produce quality output. This, however, requires substantial training and technical advice, and in some cases support to cope with higher production costs, and the initial certification costs, during the conversion period.

In order to ensure high productivity and product quality, companies or cooperatives that organize smallholders for organic production should take care that farmers have access to appropriate inputs such as seeds of suitable varieties, organic manures and pest-management means. If production costs are to be kept low, the focus should be on farm-own or locally available resources. Successful initiatives usually train farmers on which inputs to use in which situation, and stimulate them to experiment with locally available materials.

In situations where organic farms are located amidst conventional farms that use pesticides, fertilizers or GMO seeds (genetically modified organisms), the organic farming initiative needs to take precautionary measures to avoid contamination from neighbouring farms. Contamination can be in the form of surface irrigation water passing through conventional fields and thus potentially carrying fertilizers or pesticides, wind drift from spraying pesticides, or pollen carried by wind or insects from GMO to organic crops. Measures such as keeping minimum distances to conventional fields, growing buffer crops and controlling irrigation water flow are usually sufficient to prevent contamination.

In order to ensure that the produce finds a market and achieves a reasonable price and organic premium, it is crucial that the supply system is oriented towards the demands of the market. Producers and their organizations should therefore analyse in advance what products are in demand and which varieties and quality specifications are required. They further need to assess whether they can produce the products in a profitable way, whether they can meet clients' (changing) requirements concerning quality, volumes, packaging and logistics, and whether they can compete with other suppliers. The business should be based on realistic price expectations – in the long term, one should be able to make a profit with a 10–15 per cent organic export premium. Diversifying production as well as markets and thinking early of a fallback (domestic) market helps to cope with – inevitable – fluctuations in market demand and prices.

Efficient organization of farmers for the market

Smallholders can be organized in different ways in relation to the unit that is marketing their products. They can form a cooperative or a similar type of producer organization that takes the raw material from the individual members and looks after the marketing. They can also be organized and contracted by a trade house which buys and sells the product. Although producer cooperatives have an inherent focus on farmers' benefits and have therefore often been the preferred choice of development organizations, they tend to have difficulties in building up and maintaining efficient management structures and skills. Companies that relate to producers or producer groups through contract farming are often more efficient in setting up a sustainable business than when a new producer organization is formed. Provided that the company deals with the farmers in a fair and transparent way, such a private sector set-up will be more beneficial for smallholders than a poorly managed cooperative.

Producer cooperatives should be careful with handling business activities that go beyond their management capacity. A reasonable division of tasks could be that the producer organization is in charge of production, extension, internal control and bulking, and then sells the raw material to a company that covers trade finance, packaging, marketing and export. Certain functions like the provision of inputs, quality management and first-level grading can also be initially covered by the company, and then transferred to the farmer organization when its capacity is built.

Appropriate extension and training are crucial for the success of the organic production initiative. These services are usually provided by the intermediary organization – the cooperative or company that markets the produce. Farmers, however, are a heterogeneous group and therefore require adapted extension approaches that work for the different types of farmers. Building an effective extension system relies on having qualified and experienced staff who are familiar with the farmers' needs – and who stay with the organization for some time! The extension system in organic production initiatives usually also absorbs functions of an internal control system that allows for smallholder group certification at a reasonable cost. Extension and internal control systems, if well managed, can help to achieve better farm management and better product quality. A well-designed system can also enable smallholders to comply with several standards at a time (e.g. organic, fair trade, Rainforest Alliance, GAP) and thus allows supply to different markets.

Intermediaries between smallholders and the demand system will only be able to provide useful services if they are based on a viable business case. They need to be able to earn sufficient income to pay farmers an adequate price for their products, to cover the costs for extension, certification and marketing and to cover the overhead and management costs of the organization. Initially, most of these businesses will incur losses until systems and client relations are established and a sufficiently large scale of operation has been reached. This may take three to five years. The capital to cover these losses is usually invested either by companies that are involved in trading the produce, or by development organizations (grants). Some international banks offer special financial products to provide capital or trade finance for organic or fair trade producer organizations. Calculating costs and margins and handling cash flows requires a high level of financial management skills.

Developing partnerships with clients

Selling agricultural products on the spot market exposes smallholders to high market volatility and fierce competition for price. Organic farming initiatives are usually linked to vertically integrated value chains in which buyers enter into a long-term partnership with certain producer groups. Vertical integration has the advantage that routes get shorter (avoiding middlemen), controlling quality and traceability becomes easier, and supply can be directly matched with demand in terms of volumes and product specifications. Ideally, this leads to higher transparency and efficiency. In addition, the buyer gets a product with a 'story' that can be used in communication to the final consumer.

Building trustful and loyal relationships between smallholder groups and buyers, however, requires time and specific efforts. Once they are established, relationships with long-term buyers can entail much more than buying the produce at a fair price. Many buyers are ready to provide pre-financing, support to improve product quality, market know-how and sometimes even provide direct support to the community.

Appropriate support systems

In many cases, the development of organic market chains is facilitated by development NGOs and business development programmes. Their role is usually a temporary one, needed until the time when the business is economically and institutionally viable and the value chain is functioning well. Facilitators develop and strengthen the capacities of chain actors and service providers, and help overcome hurdles and bottlenecks. In a situation where organic production and marketing are entirely new, the facilitator can also make information accessible, stimulate innovation, and support interested chain actors in building the necessary capacities.

One important function of facilitation is to link suitable stakeholders (producer organizations, processors, buyers, certification agencies, finance) and to ensure that they mutually communicate their requirements and coordinate their activities. The facilitator needs to make sure that all stakeholders are heard, and mediates between the different interests of the chain actors. However, there are a number of activities that facilitators should be very careful not to get engaged in. If an NGO or government programme that is designed to last only for a few years takes up core functions of a value chain, the entire chain is likely to collapse once the support ends. A competent agricultural advisory service and an internal control system, for example, are essential for the functioning of an organic production initiative. These services should therefore not be provided by a development project. However, the facilitator can support the initial development of the necessary capacities by helping the actors to design suitable systems and tools, and to recruit and train the required staff.

Blanc and Kledal (2012) show how NGOs, faith-based organizations and public entities in Brazil were all strongly committed to the inclusion of smallholders in the organic market, lowering the many transaction costs *ex ante*, be it within the alternative agro-food system, or the formalized domestic- and export-oriented systems. However, when smallholders entered the modern commercialized agro-food systems, the array of multifaceted problems and transaction costs *ex post* were not seriously and professionally addressed. Due to the fact that NGOs and faith-based organizations were sometimes reluctant to support smallholders in commercially market-oriented food systems, Blanc and Kledal (2012) suggest that efforts should be taken to provide a policy frame, which enables public-related entities to both secure a sustainable inclusion, as well as provide exit strategies for those who experience market exclusion.

Examples of elements in such a policy frame could be (a) the establishment of a task force pro-actively helping and training farmers to develop clear principal-agent rules when setting up a cooperative or a market organization, (b) mediating and functioning as a neutral third party when crises occur in farmers' organizations, (c) facilitate horizontal flows between the different agro-food systems, thereby compensating the bias of some NGOs, and (d) provide relevant exit strategies whether related to non-farm activities, or part-time farm opportunities. In other words, the role of the state would be to establish an institutional framework supporting market transactions that can secure longer-term legitimacy of organic food production, economic growth and social stability, as opposed to the cases with short-term charity engagement on profit-oriented markets. None of them proved sustainable in the Brazilian study.

The success of an organic value-chain initiative depends to a considerable extent on the business environment in which it operates. Some aspects of the business environment of a specific country cause obstacles to agro-businesses in general. Weaknesses in transport infrastructure, financial services and legal systems affect many types of businesses, and are not easily changed. In many countries, there are national or international schemes to support the development of agri-businesses. Often they include cost-sharing arrangements for setting up processing or storage facilities, or export promotion programmes.

However, there are some aspects that cause obstacles specifically to organic businesses. Pesticide application schemes, compulsory fumigation of agricultural goods for export, fertilizer subsidies and the promotion of GMOs are typical examples. On the other side, governments are increasingly taking an interest in the development of the organic agriculture sector. Governments that want to create an enabling environment for organic businesses could formulate an organic agriculture policy.[4] Suitable elements of an organic sector policy are:

- inform farmers and companies about organic agriculture
- support the set-up of organic extension services and internal control systems
- promote recycling of agricultural waste
- promote consumer education and awareness on organic agriculture
- collect and publish data on organic production and markets
- develop national standards and regulations to foster the domestic market and reach the requirements of international markets to reduce entry costs
- facilitate development of the domestic market; encourage public procurement of organic products
- support export promotion activities, e.g. participation in trade fairs
- establish organic research and seed-breeding programmes
- include organic agriculture in the curricula of schools and universities.

Box 7.2 Agro-chain partnership – BioUganda Ltd and EPOPA (Export Promotion Of Organic Products from Africa)

BioUganda Ltd is a family-owned Ugandan company which deals in organic fresh and dried tropical fruits. The main product is fresh and dried pineapples, but it also supplies passion fruits, apple bananas, mangoes and ginger to the European market, and some to South Africa. The fruits are sourced from around 200 farmers located in different regions of Uganda to secure a stable supply over a long period. EPOPA supported BioUganda over a three-year period with assistance and capacity building of the field organization, mobilization and training of the farmers, certification issues and initial costs, the installation and operation of the drier unit, product development and packaging, financial and quality controls. Investment in a drier allowed for a diversified product range and enabled BioUganda to use second-grade and oversized fruits which were not suitable for fresh exports. This saves on the weekly logistics, management and the promotion of the products and securing of markets. Payback period (value of investment balanced with extra income for the smallholders) for the BioUganda investment is expected to be eight years. Every year BioUganda exhibits its products in the African Pavilion at the Biofach, the largest organic trade show.

Source: www.epopa.info

Concluding summary

Increasingly, national governments and international development agencies view commercialization and inclusion of smallholder farmers into modern value chains as a central policy in achieving poverty reduction. Organic food and farming, due to the sector's fast growth worldwide, is in this respect becoming an important part of national pro-poor agricultural policies in developing and emerging countries. However, poor smallholders, from subsistence to small-scale commercial farmers, face numerous types of constraint as well as different transaction costs, when trying to enter commercial value chains.

First, the types of constraint smallholders endure are strongly related to the agro-food system and the value-chain organization it plans to operate within, be it:

*   alternative local markets
*   formalized urban domestic markets
*   formalized modern export markets.

Logically a subsistence farmer entering the high quality requirements of the export-oriented agro-food system will face a whole range of different challenges than a semi-commercial farmer joining a modern domestic-oriented procurement system.

Second, an analysis of the constraints along an agro-food value chain should be subdivided and classified whether they belong to:

- the supply system
- the intermediaries
- the demand system
- the supporting resource environment.

Each of these entities holds different types of constraints and hence specific transaction costs. The major challenge is then how to coordinate information flows and trust building between the actors of the value chain both *ex ante*, during a transaction and *ex post* in order to minimize transaction costs and hence secure sustainable profits (value) along the whole chain. In other words, certain institutions ('rules of the game') need to be put in place, which, in a development perspective, will often require a strong committed supporting resource environment focusing on longer-term capacity building as well as securing sustainable agro-chain partnerships.

By supporting the development of organic value chains that link smallholders to markets, governments, donors and development agencies can contribute to more sustainable resource management, better livelihoods of the involved farmers and workers, and more employment and value generation in the producing country. Not all organic initiatives, however, automatically result in viable value chains that can run on their own once the support ends. In some cases, donor intervention may even hinder promising organic businesses from flourishing, as it can hamper emerging entrepreneurial thinking, subsidize competition and distort the market. The long-term effect of the intervention largely depends on how the support programme is designed. Support programmes especially need to have an exit strategy to ensure that by the end of the project intervention businesses established are institutionally and economically sustainable.

Notes

1   In economics the 'hold-up problem' is a situation where two parties (such as a supplier and a manufacturer or the owner of capital and workers) may be able to work most efficiently by cooperating, but refrain from doing so due to concerns that they may give the other party increased bargaining power and thereby reduce their own profits. It has been described as the 'most influential work' in recent decades on why firms exist and what determines their boundaries. Hold-up situations are common in agriculture where farmers produce perishable products that need to be sold soon after harvesting like fruits and vegetables. If the buyers downstream have a strong bargaining power they can put the farmer in a hold-up situation where the farmer has to sell his/her products at a much lower price. The potential hold-up situation in farming is an important explanation for why farmers often are seen entering into cooperatives or market associations.
2   The idea of four agro-food systems, when working with organic food and non-food markets in developing countries, is building on the work of McCullough et al.,

2008. However, these authors only operate with three conventional food systems prevalent in developing countries.

3 For a comprehensive description on how to set up and manage organic production initiatives with smallholders, see *The Organic Business Guide* (Elzakker and Eyhorn 2010)

4 See *Best Practices for Organic Policy* (UNEP–UNCTAD 2008)

References

Barrett, H.R., Browne, A.W., Harris, P.J.C. and Cadoret, K. (2001) 'Smallholder farmers and organic certification: accessing the EU market from the developing world', *Biological Agriculture and Horticulture*, 19, 2, pp. 183–199

Blanc, J. (2009) 'Family farmers and major retail chains in the Brazilian organic sector: assessing new development pathways. A case study in a peri-urban district of São Paulo', *Journal of Rural Studies*, 25, 3, pp. 322–332

Blanc, J. and Kledal, P.R. (2012) 'The organic sector of Brazil: prospects and constraints of facilitating the inclusion of smallholders', *Journal of Rural Studies*, 28, 1, pp. 142–154

Bolwig, S., Gibbon, P. and Jones, S. (2009) 'The economics of smallholder organic contract farming in tropical Africa', *World Development*, 37, 6, pp. 1094–1104

Coase, R.H. (1937) 'The nature of the firm', *Economica*, 4, pp. 386–405

Crucefix, D. (1998) 'Organic agriculture and sustainable rural livelihoods in developing countries', study commissioned by the Natural Resources and Ethical Trade Programme (Natural Resources Institute) and conducted by the Soil Association in the context of the Department for International Development Natural Resources Advisors Conference in July

Eenhoorn, H. and Becx, G. (2009) 'Constrain constraints! A study into real and perceived constraints and opportunities for the development of smallholder farmers in Sub-Sahara Africa', Wageningen University and Research Centre Wageningen, available at http://www.wur.nl/NR/rdonlyres/453186A9-2388-402A-B616-D9F8 EDB91BEC/79143/EenhoornBecxfinalConstrainConstraints.pdf

Egelyng, H. (2009) 'Organic agriculture: glocalisation options for the South?', in R. Janrdhan and A.S. Sisodhya (eds), *Organic Farming – Perspectives and Experiences*. ICFAI University Press, Bangalore, pp. 186–201

Elzakker, B. and Eyhorn, F. (2010) 'The Organic Business Guide – developing sustainable value chains with smallholders'. IFOAM and collaborating organizations (Helvetas, Agro Eco Louis Bolk Institute, Icco, UNEP), available at: www.unep. ch/etb/publications/Organic%20Agriculture/Organic%20Business%20Guide%20 publication/Organic_Business_Guide_Eng.pdf

Gomez-Tavar, L., Martin, L., Gómez Cruz, M.A., and Mutersbaugh, T. (2005) 'Certified organic agriculture in Mexico: market connections and certification practices in large and small producers', *Journal of Rural Studies*, 21, 4, pp. 461–474

Gonzalez, A. and Nigh, R. (2005) 'Smallholder participation and certification of organic farm products in Mexico', *Journal of Rural Studies*, 21, 4, pp. 449–460

Hobbs, J.E. (1997) 'Measuring the importance of transaction costs in cattle marketing', *American Journal of Agricultural Economics*, 79, pp. 1083–1095

IFAD (International Fund for Agricultural Development) (2002) 'Thematic evaluation of organic agriculture in latin America and in the Caribbean'. Evaluation Committee

Thirty-second Session, Rome, 9 December. Available at: www.ifad.org/gbdocs/eb/ec/e/32/EC-2002-32-W-P-3.pdf

IFAD (International Fund for Agricultural Development) (2005) 'Organic agriculture and poverty reduction in Asia: China and India focus. Thematic evaluation', July, Report no. 1664

Källander, I. and Rundgren, G. (2008) 'Building sustainable organic sectors'. IFOAM January. Available at: www.ifoam.org/pdfs/Building_Sustainable_Organic_Sectors_WEB.pdf

Kilcher, L. (2007) 'How organic agriculture contributes to sustainable development', in C. Hülsebusch, F. Vichern, H. Hemann and P. Wolff (eds), 'Organic Agriculture in the Tropics and Subtropics – Current status and perspectives'; supplement 89 of the *Journal of Agriculture and Rural Development in the Tropics and Subtropics*, Kassel University Press, pp. 31–49

Kledal, P.R. (2009) 'The four food systems in developing countries and the challenges of modern supply chain inclusion for organic small-holders', conference paper for IRN (International Rural Network) conference August, Udaipur, India

Kledal, P., Hui, Q.Y., Egelyng, H., Yunguan, X., Halberg, N. and Xianjun, L. (2007), 'Organic food and farming in China', in Willer, H. and Yussef, M. (eds) *The World of Organic Agriculture – statistics and emerging trends 2007*. Bonn and Frick: IFOAM and FIBL, pp. 114–119

Kledal, P.R., El-Araby, A. and Salem, S.G. (2008) 'Organic Food and Farming in Egypt', in Willer, H., Yussefi-Menzles, M. and Sorenson, N. (eds) *The World of Organic Agriculture – statistics and emerging trends 2008*, Bonn and Frick: IFOAM and Fibl, pp. 160–163

Kledal, P.R., Oyiera, H.F., Njoroge, J.W. and Kiarii, E. (2009) 'Organic food and farming in Kenya', in *The World of Organic Agriculture – statistics and emerging trends 2009*, IFOAM, Bonn and FiBL, Switzerland, pp. 127–132

McCullough, E.B., Pingali, P.L. and Stamoulis K.G. (eds) (2008) 'The transformation of agri-food systems – globalization, supply chains and smallholder farmers', FAO

Masten, S.E. (2000) 'Transaction cost economics and the organization of agricultural transactions' in M.R. Baye and B. Elwert (eds), *Industrial Organization*, Elsevier Science, New York.

Nordlund, E. and Egelyng, H. (2008) 'Perceived constraints and opportunities for Brazilian smallholders going organic: a case of coffee in the state of Minas Gerais', in *Cultivating the Future Based on Science, Proceedings from the Second Conference of the International Society of Organic Agriculture Research (ISOFAR), Modena, Italy, June 18–20*, pp. 462–465

Pingali, P.L. and Rosegrant, M.W. (1995) 'Agricultural commercialization and diversification: processes and policies', *Food Policy*, 20, 3, pp. 171–185

Pingali, P.L., Khwaja, Y and Meijer, M. (2007) 'The role of the public and private sectors in commercializing small farms and reducing transaction costs', in J.F.M. Swinnen (ed.), *Global Supply Chains, Standards and the Poor – How the Globalization of Food Systems and Standards Affects Rural Development and Poverty*, CABI International, Wallingford

Reardon, T. and Berdequé, J.A. (2003) 'The rapid rise of supermarkets in Latin America: challenges and opportunities for development', *Policy Review*, 20, 4, pp. 371–388

Scialabba, N. (2007) 'Organic agriculture and food security'. OFS/2007/5, Food and Agriculture Organization of the United Nations (FAO), Rome, Italy. Available at: www.fao.org/organicag/ofs/docs_en.htm.FAO, 2003

Shah, T., Verma, S., Bhamoriya, V., Ghosh, S. and Sakthivadivel, R. (2005) 'Social impact of technical innovations. Study of organic cotton and low cost drip irrigation in the agrarian economy of West Nimar Region', Report for the Research Institute of Organic Agriculture (FiBL), Research Program 'Organic Cotton Farming in India'

Siriex, L., Kledal, P.R. and Sulitang, T. (2011) 'Organic food consumers' trade-offs between local and imported, conventional or organic products: a qualitative study in Shanghai', *International Journal of Consumer Studies*, 35, 6, pp. 670–678

UNEP–UNCTAD (2008) 'Best practices for organic policy – what developing country governments can do to promote the organic agricultural sector', Available at: www.unepunctad.org/cbtf/publications/Best_Practices_UNCTAD_DITC_TED_2007_3.pdf

Willer, H. and Yussefi, M. (2006) *The World of Organic Agriculture – statistics and emerging trends 2006*, FiBL-IFOAM Report. Bonn and Frick: IFOAM and FiBL

Willer, H. and Kilcher, L. (eds) (2011) *The World of Organic Agriculture – statistics and emerging trends 2011*, FiBL-IFOAM Report. Bonn and Frick: IFOAM and FiBL

Williamson, O.E. (2000) 'The new institutional economics: taking stock, looking ahead', *Journal of Economic Literature*, 38, 3, pp. 595–613

Case study 4
Certified organic and fair trade impacts on smallholders' livelihoods in Kandy case area, Sri Lanka

S. Vaheesan, K. Zoysa and Niels Halberg

The organic agriculture in Kandy area (Sri Lanka) was initiated in the 1990s by two different pathways: the NGO Gami Seva Sevana (GSS) started credit and savings groups among poor smallholder farmers in 1994 and insisted that new members converted to organic agriculture and diversified their small land-holdings. Farmers in this area had neglected tea production because of low prices but GSS encouraged organic tea cultivation. The GSS has kept a holistic household approach to the development support and offered a number of training courses and extension services to the members in order to empower women and men. In 1993 another initiative in organic tea production was launched as a combination of small farmers' organizations (SOFA) and a processing and trading company, Bio Foods, which specializes in tea and spices purchased from small-holder farmers and exported. Since 2002 the village-based SOFA groups have been formally independent of Bio Foods though farmers supply all their organic tea to the company. While the organic price premium is paid directly to individual farmers the fair trade premium is channelled to the SOFA farmer groups via the head office. The premium consists of 40 rupees per kg tea, which amounts to approximately 3 million rupees every year. These funds are used mainly for community-level support such as tree planting and for non-agricultural support to families such as scholarships for students. Together, Biofoods and SOFA organize training of farmers (including specialized courses for women). Not surprisingly the Bio Foods company has had a stronger focus on the marketing of the products (tea and spices) compared with GSS. Thus, in order to benefit from the stronger market links through Bio Foods the GSS farmer groups recently transformed into associations and became suppliers to the Bio Foods factory.

The company does not operate with written contracts to secure the supply from farmers but has a business relationship building on mutual trust and confidence and on long-standing commitment to purchase the entire production of the attached farmers at the guaranteed price. The SOFA is comprised of about 27 'societies' where each society is composed of several small village-based groups of farmers (Figure 1). In all the farmer groups comprise a total of more than 1,800 member farmers coming from different organic tea growing areas in Sri Lanka, mainly from central regions.

Figure 1 Members of a SOFA group in front of their offices and meeting facilities

Certified organic farming and small farmers' livelihood impacts

The qualitative interviews with different stakeholder groups revealed that historically the organic projects had targeted relatively resource-poor farmers, whose livelihood was mostly subsistence farming, supplemented with very poorly paid hired labour in tea plantations. The GSS and Bio Foods successfully managed to support the farmers in improving their lands for their cash crop production while at the same time establishing market links for certified products with a price premium. The farmers who participated from the beginning or shortly after were quite unanimous in their praise of the initiative's success in terms of changing their livelihood through improved land use and income, local employment opportunities related to the project and higher self-esteem and capacity as smallholder farmers.

The average value of crops sold was slightly lower in the organic farms compared with non-organic even with a price premium of nearly 100 per cent. This was caused by significantly lower yields, which was explained by the local experts by the fact that the conventional tea gardens had already been restored with new vegetative propagated (VP), high-yielding varieties instead of the seedling tea still used in organic gardens (which gives less yield in the form of green leaves). They expected that once the entire tea population has been replaced with VP tea in the organic tea lands, the yield from organic tea would increase and the yield difference would become smaller. However, the variable costs including own labour costs were significantly lower in the organic system compared with the non-organic even though costs of own labour were higher in the organic system. In combination, this resulted in a comparable level of net profit from crops in the two systems after payment of own labour. Thus, the organic system with a lower investment in terms of cash inputs and higher input of own labour was a competitive method of gaining profit from the forest tea gardens.

The total budget prepared by SOFA for using the fair trade premium in 2005 was 5.4 million rupees, which was shared among the 27 village-based SOFA groups. This amounts to 3,000 rupees per member but the money is seldom distributed to individual households. Most of the funds are given for community development projects proposed by village SOFAs and approved by the General Assembly. Two main areas targeted for support in 2005 – besides infrastructure development within SOFA itself – related to agricultural development. Other areas supported were capacity building for women of the member families and general welfare, including scholarships for the children of farmers for further education. The effect of the SOFA groups in terms of community development and building of social capital and organizational capacity is just as important as the amount of funds distributed. A key activity of SOFA has been to support women's self-employment (Figure 2) and there is an increased focus on involving women in training programmes supported by fair trade funds. This is linked with the increased possibilities for women gaining an income from rural work either on their own land or working as hired labour in organic cash crop production. The respondents appreciated the job opportunities for women near their residential villages created by the market for organic cash crops. On the other hand, the Kandy case area shows that the improvement in the wider dimension of social capital – which is linked to participation in and capacity for grassroots organization – primarily should be ascribed to the rules for fair trade certification rather than the organic certification.

Figure 2 A SOFA group treasurer demonstrating reed boxes for marketing tea and payment scheme for female self-employment

The Fairtrade Labelling Organization (FLO) rules that demand the establishment of farmer-led local organizations or cooperatives to collectively manage the funds from the price premium enhance the potential to build social capital and improve women's involvement in community-level decision making.

Other important non-monetary benefits of organic agriculture include reduced health problems and risks from pesticide use with a lower incidence of organic households reporting pesticide-related health problems and accidents. More than 25 per cent of non-organic households surveyed reported that a family member had needed medical attention due to pesticide poisoning and the majority knew of local pesticide accidents. However, during qualitative interviews it was stressed that the non-use of pesticides by the organic farmers was spreading to non-certified farmers partly because of organic farmers' influence and the risk of pesticides drifting from one farm onto a neighbouring organic farm and this being caught in tests. Thus, in this case of smallholder tea production, use of pesticides and subsequent contamination was now not a major issue among the conventional farmers.

Organic farming and soil-fertility improvement in Kandy case area

When converting to organic tea cultivation, the main concern in the tea small-holder lands has been to re-establish the degraded lands and rejuvenate the soil fertility to its potential productive capacity. The low soil fertility due to heavy erosion and scarcity of organic materials for incorporation into the fields posed major problems for the organic production. The use of manure is modest due to very little livestock keeping and composting is apparently not sufficient in its present form. Since the availability of manure in the area is limited, farmers bring manure from nearby districts. Hence, many organic farmers have reported lack of availability of manure as a constraint for adapting organic methods of fertility management. Some of the SOFA groups have applied their Fair Trade premiums to set up local herds of small ruminants with the (additional) purpose of getting more manure. Moreover, a number of agro-ecological techniques have been introduced such as composting, mulching, using insect trap crops and intercropping besides the overall idea of promoting the so-called forest garden tea system, where tea is grown in a mixture of trees, palms and spices (Figure 3). Organic farmers constructed stone hedges (stone terraces) along the contour lines of sloping land to curb soil erosion before (re-)planting the tea bushes. The extension officers of the GSS or Bio Foods trained farmers in many practices

Figure 3 Farmer in a sloping forest tea garden with mixtures of spices

for soil conservation. They demonstrated that the medicinal plant savandera (*Vetiver zizanoides*) could be used to establish live hedgerows in the lower section of the stone terraces. Further, extension officers directed the farmers to plant Albizzia (*Albizzia lebbek*) and Gliricidia (*Gliricidia maculata*) as shade trees between the tea bushes. Both Albizzia and Gliricidia used as medium shade trees are members of the plant family Leguminosae that fixes nitrogen in the soil; they shed much of their leaf litter and hence increase the biomass and fertility of the soil.

Much attention was paid initially to spur the rate of growth of the tea as well as increasing the soil fertility. Weeds were removed by hand and the removed weeds were left as mulch on the soil floor. Farmer participatory research conducted by the Tea Research Institute of Sri Lanka in collaboration with GSS in farmers' fields showed that compost application followed by foliar spraying of diluted Gliricidia leaf extract on tea foliage has improved the yields significantly. Efficient recycling through composting is mainly implemented by those who rear animals. Without animal manure, composting is less effective and less accepted as a practice by the growers, hence they usually built compost heaps in the vicinity of animal sheds. However, only a few farmers produce sufficient quantities to manure their tea bushes with compost. The compost is spread underneath the bushes; sometimes quantities do not exceed a handful per bush. After pruning, some growers bury pruning materials in 15–20-cm deep trenches between tea rows together with compost and animal dung once in a pruning cycle of 3–4 years.

Conclusion

Smallholder farmers improved their income (due to a combination of price premium and cost reduction) and livelihood through their participation in the production and marketing of certified organic and fair trade products organized by private companies and NGOs. Organic households integrated a larger diversity of crops and livestock, giving a potentially more diverse diet, than non-organic households. Organic farmers improved soil fertility and land use by adopting a number of agro-ecological and soil conservation methods. Moreover, adoption of organic farming resulted in reduced health problems and pesticide poisoning. The participation in certified organic and fair trade largely benefited local empowerment at the village level and increased organizational capacity, which again supported community development including infrastructure development, scholarships for students and capacity building for women.

8 Policies and actions to support organic agriculture

Sophia Twarog

Introduction

Despite overwhelming evidence on the wide range of benefits of organic agriculture and other forms of ecological/sustainable agriculture, few governments are actively and adequately supporting the development of this form of agriculture in their countries. This chapter addresses three key questions that need to be answered in light of this situation:

- Why should governments engage in supporting sustainable forms of agriculture?
- What are the challenges to a change towards more sustainable agriculture?
- What should governments do to support sustainable agriculture?

Why should governments care?

Governments have limited resources and must be selective about where and how they use them – both financial resources as well as human resources in terms of time and attention given to matters. There is a very good case for government action to promote the development of sustainable agricultural production systems, including organic agriculture systems. Key reasons for action include:

- responding to changes in what their constituencies want
- correction of massive market failure
- coping with future big global challenges
- the promise to meet international commitments including the Millennium Development Goals (MDGs).

Changes in what people want

Populations in general are becoming much more conscious of health and environmental issues. Companies advertise how green they are. Consumers are more interested in knowing where their food, fibre, wood and other products are coming from, how they are produced, and how resource efficient they are.

They are worried about the health consequences of industrial agriculture, ingesting pesticides, risks of food scares like E. coli, salmonella, dioxins, mad cow disease and bird flu. Most do not feel good about factory farming, the harsh realities of which are still kept discreetly away from shoppers at the meat aisle at the supermarket.

People increasingly want to buy safe healthy food that has been produced in an environmentally and animal welfare-friendly manner. This includes organic products. Markets for certified organic products have tripled since 2000, reaching US$ 55 billion in 2009. Even in the recent economic downturn, markets for organic products continued to grow (Sahota, 2011). Moreover, production and sales of uncertified organic (or near-organic) products are likely 10 to 20 times greater. Many sales take place on local markets with local relationships, knowledge and trust filling the place of third-party certification.

Nearly all consumers given an informed choice between equally or near-equally priced agro-industrial products and organic products would choose the organic product. The smaller supply and higher retail prices of organic products act as a deterrent. These are in turn the result of market failure.

Market failure

Agriculture is a classic case of market failure due to externalities. The public costs of the current system from environmental damage and health costs are not borne fully by the private decision-makers and asset holders within the sector. The costs are largely borne by society as a whole (UNEP, 2011). Thus with the current incentive structure, private decision-makers will not make choices that are in line with the optimal solution from a societal point of view. Government intervention and action are necessary to correct such market failure.

The externalities from the current predominant agricultural production system accrue in many contexts. First, there is overuse of resources including water, energy and soil fertility: agriculture currently absorbs 70 per cent of the world's freshwater withdrawals (Turner et al., 2004). For example, it has been estimated that in the United States, producing 1 kg of wheat uses 900 litres of water and 1 kg of beef uses 100,000 litres (Pimental, 1997). In addition, industrial agricultural systems use more energy than they produce in terms of calories. The energy input is hidden, for example, in the production of agro-chemicals. Agro-industrial systems consume on average 10 energy calories for each food calorie produced (UNEP, 2011). Industrial production systems tend to focus on short-term productivity gains rather than on long-term sustainability. This attitude leads to reduction of innate soil fertility, mining its nutrients and destroying the complex web of life in the soil (Matson et al., 1997). The soil becomes deprived of nutrients, a simple substrate to which necessary ingredients for growing something are added. Degraded soils also retain less water. This can lead to lower groundwater levels and water stress.

Second, current agricultural practices lead to considerable pollution of soil, air and water: they lead to a build-up of toxic pesticides and herbicides in water and soil. Run-off of fertilizer into water systems creates blooms of oxygen-depleting algae that kill fish and disrupt aquatic ecosystems. The European Nitrogen Assessment (ENA) concluded that the overall environmental costs of nitrogen in agriculture outweigh the economic benefits (Sutton et al., 2011). In addition, there is air and water pollution caused by production and distribution of agro-chemicals. Finally, industrial agriculture causes enormous waste management and soil-fertility problems due to the spatial separation of livestock and crop production (Tilman et al., 2002; Steinfeld et al., 2006).

Third, climate change is an issue. Agriculture directly contributes about 10 per cent to global greenhouse gas emissions. Adding contributions from land-use change and indirect emissions from production inputs, this share rises up to 30 per cent (Smith et al., 2007, 2008; Bellarby et al., 2008).

Fourth, the current agricultural production system has led to a considerable reduction in biodiversity. Industrial agriculture reduces biodiversity in the soil, on the farm and surrounding the farm. There are drastic reductions in the whole range of species including micro-organisms, earthworms, butterflies, bees, birds, mammals and more. Over 70 per cent of the world's agro-biodiversity has been lost due to the standardization convenient for the prevailing agro-industrial model's features, including mechanization, monocropping and control of input markets. Traditional varieties and the associated knowledge about their cultivation are being lost forever. This puts the whole food system at risk as it reduces the genetic reservoirs available in case of emergencies (Twarog and Kapoor, 2004).

Fifth, besides these environmental impacts, there are direct impacts of prevalent agricultural practices on individual human health. Agro-chemicals are a leading cause of occupational morbidity and mortality worldwide. In developing countries, up to 14 per cent of all occupational injuries in the agricultural sector and 10 per cent of all fatal injuries can be attributed to pesticides (Myers, 2000). Agro-chemicals cause illness and death not only to farmers and farm workers, but also for populations living near or downstream from farms and households where these chemicals are used. These effects are difficult to track and therefore generally ignored. There are many indications of these negative effects though, and significant increases in cancer, Parkinson's disesase and stillbirths have been observed (Chemtox, 2011).

Furthermore, agro-chemicals consumed in food have a known negative health impact. Most pesticides and herbicides are carcinogenic. A simple rinse in cold water does not guarantee their complete removal. Some penetrate the skin of the product. In animal food, some chemicals become accumulated along the food chain. All this and the illness and deaths of farm workers and their families are not reflected in the price of the agricultural products.

Big global challenges ahead

From this short assessment, the consequences of inaction are clear. Furthermore, new challenges are on the horizon. These include climate change, rising energy prices and peak phosphorus.

The lack of cooperation and coordinated swift action to combat climate change has destined the globe to a rise in average temperature and increased climatic variability. The sector that will be most directly and immediately affected by this is agriculture. The rains will come at unpredictable times of year, if they come at all. Or they will come, but in torrents that wash away the nutrient-rich topsoil. Crops that were successfully grown in a region will fail. New loci of production must be identified. The knowledge farmers have about certain crops and livestock will become irrelevant as the agro-ecological zone shifts literally beneath their feet (Easterling et al., 2007). High adaptive capacity will be the key to future agricultural success (cf. Chapter 5). Sustainable agricultural production systems, including organic agriculture, are based on building strong and fertile soils. These production systems are much more resilient in the face of climatic stress. These fields stay green longer in times of drought and hold on to more nutrient-rich topsoil when it pours (Twarog, 2006). These systems are also knowledge- and skills-intensive. Organic agriculture develops the human and social capital of farmers and rural communities (UNEP-UNCTAD, 2008b). These farmers and communities are much better prepared to meet challenges with innovation and joint action.

Rising prices of oil and other fossil fuels repeatedly promote the debate about just when the oil will run out. But there is no doubt that it is a finite resource and like all finite resources will have an end. Before resources run out, they become more expensive. As the 'low-hanging fruits' have been increasingly harvested, oil is now sourced from second-best sources such as oil shale. The cost of production is higher. While economic downturns, speculation and other factors lead to short-term fluctuations, in the long term prices for oil and other fossil fuels are on the rise. See Figure 8.1.

The rise in the price of oil has a huge impact on the agricultural sector, particularly the agro-industrial model. Synthetic agro-chemicals use a lot of fossil fuels in their production and transportation. For example, natural gas accounts for 70–90 per cent of the cost of production of ammonia, the primary component of synthetic nitrogen fertilizer (Sawyer, 2003). The correlation between the price of nitrogen fertilizer and natural gas has been estimated at 0.74 (Agriculture and Agri-Food Canada, 2010). This has important consequences because the higher the price of fossil fuels, the less competitive the agro-industrial model and the more competitive the sustainable/organic agriculture model. Higher fuel prices thus have the same effect on agricultural producers as an internalization of external costs.

Another problem looming on the horizon is peak phosphorous. Industrial agriculture relies on mineral fertilizers, the well-known N (nitrogen), P (phosphorous) and K (potash). These cannot be substituted by other substances. Almost all phosphorus used today – and all phosphorus in chemical fertilizers in particular – comes from non-renewable resources. Estimates of the time when global phosphorus deposits will be depleted vary slightly, but indications are that there will not be sufficient phosphorous to meet agricultural demand within 30–40 years (Elser and White, 2010). The timing of peak phosphorous is related

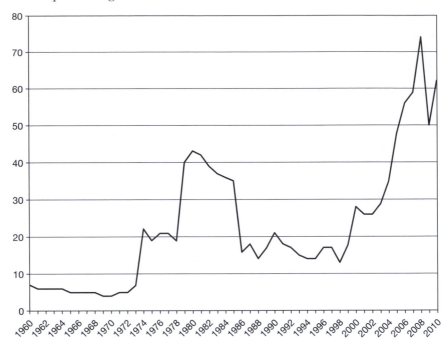

Figure 8.1 Price of oil 1960–2011

Note: unit: constant 2000 $
Source: World Bank Commodity Price Data, 2011.

to that of peak oil. Higher oil prices increase demand for biofuels which, as currently produced, in turn increase demand for phosphorous (White and Cordell, 2010).

As with oil, before running out, phosphorus will become more expensive. Prices are already on a long-term rise (Figure 8.2). Note the price spike just before the food price crisis in 2008.

Moreover, geopolitical aspects of phosphorous supply are a cause for concern. Some 87 per cent of known global phosphate reserves are located in just five countries: Morocco (35 per cent), China (23 per cent), South Africa (9 per cent), Jordan (9 per cent) and the United States (7 per cent) (US Geological Survey, 2010). The rest of the world depends upon imports. Even the United States has become a net importer. Some of Morocco's reserves are in the much-disputed Western Sahara territory. In some months in 2008, China safeguarded its reserves by imposing a 35 per cent tariff on exports, effectively eliminating them (Zengxin, 2008).

The current agro-industrial system based upon mined and, for most countries, imported phosphorous will face very serious problems in the coming years. Governments would be well advised to address this problem now, before the crisis hits. The solution is sustainable agricultural practices, which recycle nutrients back into the soil combined with ecological sanitation, which recycles

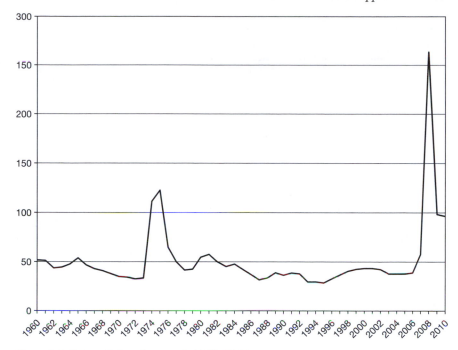

Figure 8.2 Price of phosphate rock 1960–2011

Note: unit: constant 2000 $
Source: World Bank Commodity Price Data, 2011.

nutrients from human waste. Organic standards and regulations should embrace the latter (Soil Association, 2010).

International commitments including the Millennium Development Goals

Governments have made a number of international commitments, including the Millennium Development Goals (MDGs). Organic agriculture can play an important role in meeting a number of these, including MDG 1 on poverty and hunger eradication, MDG 7 on environmental sustainability and MDG 8 on global partnership for development (see Jimenez, 2006; Setboonsarng, 2006). Organic and other forms of ecological agriculture are also a key approach to meet all three goals of sustainable development as declared at the Rio Earth Summit in 1992: economic, social and environmental sustainability.

Furthermore, organic agriculture can help meet goals under other multilateral environmental agreements. In 2010, the parties to the Convention on Biological Diversity (CBD, 2010) adopted a Strategic Plan for Biodiversity 2011–20, including the Aichi Biodiversity Targets. Several of these targets are directly related to organic agriculture in as much as they likely would be met in a context of organic production. See Table 8.1.

Table 8.1 Selected Aichi Biodiversity Targets related to organic agriculture

Target 7	By 2020, areas under agriculture, aquaculture and forestry are managed sustainably, ensuring conservation of biodiversity.
Target 8	By 2020, pollution, including from excess nutrients, has been brought to levels that are not detrimental to ecosystem function and biodiversity.
Target 13	By 2020, the genetic diversity of cultivated plants and farmed and domesticated animals and of wild relatives, including other socio-economically as well as culturally valuable species, is maintained, and strategies have been developed and implemented for minimizing genetic erosion and safeguarding their genetic diversity.

Source: CBD, 2010.

Sustainable agriculture can also help meet the goals related to the protection of wetlands (Ramsar Convention), prevention of desertification (United Nations Convention to Combat Desertification – UNCCD) and reduction of persistent organic pollutants (POP) and toxic waste (Basel Convention).

Finally, organic agriculture also has enormous potential for use in the adaptation and mitigation of climate change, thus contributing to meeting the Kyoto Targets or any reduction goals of global successor agreements, from 2012 onwards. Accounting for all indirect or grey emissions, the agricultural sector is the top GHG emitting sector in the world (about 30 per cent). The biggest mitigation potential in agriculture lies in soil carbon sequestration, which also improves soil and thus contributes to adaptation (cf. Chapter 5). This potential is realized through agricultural practices common in organic agriculture, such as optimized crop rotations with legume leys and the use of organic fertilizers. Conversion to organic agriculture production could therefore contribute to reducing global GHG emissions. Some optimized agricultural mixed–crop and livestock systems or agro-forestry projects can even be climate neutral (e.g. Soussana et al., 2010). The United Nations Framework Convention on Climate Change (UNFCCC) and other institutions of international climate policy have yet to fully recognize this potential.

Challenges to change

The two biggest challenges to change are mindsets and powerful vested interests.

Conventional mindsets

The agro-industrial model became popular after the Second World War. In the past 60 or so years, we as a society have come to believe that this is the only way to produce our food. Any other way is often considered non-scientific, non-modern. Those who have taken any agricultural training have generally been trained only in this type of agriculture. Ecological agricultural production systems have largely been looked down upon as the province of treehuggers and econuts who obviously know nothing compared to those clad in shining white lab coats armed with test tubes.

It is a case of supreme and unfortunate ignorance. In our efforts to embrace the future, we have forgotten and disregarded much of what we already knew. And we have missed the opportunity also to build upon that knowledge by applying modern scientific methods to better understand natural processes and how they can be harnessed for sustainable food production and to develop corresponding methods and practices. It is difficult to combat myths, such as the backward image of organic agriculture, especially when they are supported by powerful vested interests.

Powerful vested interests

There are strong vested interests in the conventional agro–industrial model. The input markets are controlled by a handful of companies. In 2004, six large companies[1] produced 77 per cent of the global agro-chemicals. Market concentration has been increasing. The concentration ratio CR4 (measuring share of top four companies in total) rose from 47 per cent in 1997 to 60 per cent in 2004. CR4 in seeds rose from 23 per cent to 33 per cent in the same period (World Bank, 2008, in Hoffmann, 2011).

Small groups of well-organized and -financed operators with narrowly defined interests will always have much more power than large disparate groups with less closely defined interests.[2] In this case, six companies narrowly focused on maximizing profits have had greatly disproportionate influence compared to the billions of people whose lives may be affected by negative externalities in the agricultural system due to the health impact of agro-chemicals, water pollution, flooding or drought due to soil degradation, etc. Their voices are virtually unheard.

To maximize their profits, the input companies aim to sell their products and, as much as possible, to keep out others selling similar products. To do this they influence their governments, for example, to set up and spread around the world strong intellectual property regimes and plant breeders' rights. They invest funds to file intellectual property rights (IPRs) on their products and take action against entities that infringe (or seem to infringe) these. The case of Monsanto versus Percy Schmeiser illustrates this (Schmeiser, 2011; Monsanto, 2011). Monsanto sued the organic farmer for growing Roundup Ready canola that drifted into his field. Mr Schmeiser refused to pay the fees Monsanto demanded. After years of legal battles and staggering (for an individual farmer) costs, the Supreme Court of Canada sided with Mr Schmeiser. Yet the company continues these tactics with countless numbers of farmers around the world, few of whom are willing, understandably, to face bankruptcy and extreme stress over standing up to a major corporation (Center for Food Safety, 2005).

Organic agriculture production systems are focused not on the purchase of external inputs but rather on the utilization of locally available renewable resources and nutrient recycling. The organic farmer is aiming to become as self-reliant as possible. This is in direct contrast to the interests of the companies whose profits are dependent on selling inputs to the farmers. Is it therefore not surprising that the agro-industrial entities make continuous statements undermining the validity of organic agriculture production systems.

Market concentration has also been rising rapidly on the retail side of the equation. Supermarkets now account for more than half of global packaged food sales (USDA, 2011). Supermarkets use centralized purchasing systems that favour large-scale suppliers able to meet the continuous year-round supply and stringent quality demands. Smallholder farmers often get squeezed out of the value chain (see Twarog, 2008).

What should governments do?

The responsibility of government is to act in the interest of the entire constituency. Governments can set framework and incentive structures, implement supportive policies and regulate markets. They can also provide information and bring stakeholders together. Finally, they can directly act by providing resources or purchasing products. These are a few possibilities for governmental action.

Set framework and incentive structures

The main role of governments is to correct the underlying framework and incentive structure so that public and private costs and benefits become more closely aligned. The ultimate goal is to internalize the externalities: those causing environmental, health or economic damage to others should pay for that; those providing benefits to others should be rewarded. This is the polluter pays principle.

Governments should set frameworks that give incentives to individual actors to undertake actions that have positive spillover effects on society and to desist from the opposite. They should take steps so that the price of a good reflects its full societal cost. This can be done, for example, through taxes, subsidies, standards and legislation.

The first and most important step in this direction is to remove perverse subsidies. For example, subsidizing or handing out synthetic agro-chemicals for free leads to overuse of these inputs. Moreover, such subsidies pose a serious tax on production systems that do not rely on these production inputs, negatively affecting the competitiveness of sustainable agriculture compared to agro-industrial agriculture.

For governments that wish to provide financial support to farmers, this playing field could be levelled, for example, by providing an equal amount of support per farmer or per hectare. A voucher system, for example, could be set up. These vouchers could be used for purchase of external inputs (synthetic agro-chemicals or biofertilizers and natural pest control measures such as beneficial insects) or for training on ecological agricultural production techniques, for example, on building soil fertility, erosion control, composting, effective intercropping and crop rotations. Such a system would equally promote the acquisition of knowledge to become a better farmer and the purchase of external inputs. Knowledge is a non-perishable, regenerative and transmissible product.

For most developing countries this would have the further added impact of improving the balance of trade. In Sub-Saharan Africa, for example, over 95 per cent of all synthetic agro-chemicals are imported from abroad representing a huge portion of these countries' import bills.[3] Training in sustainable agricultural production techniques, on the other hand, can lead to the same increases in yields and even higher incomes for the farmers, and can be given by local service providers complemented by occasional training of trainers and influxes of new knowledge and best practices from abroad.

Putting a floor on the price of oil could also shift agricultural production in a more sustainable direction. Oil is deeply embedded in the price of synthetic agro-chemicals, as it is used in their production and transport. The higher the price of oil, the higher the price of these inputs, the less competitive the agro-industrial production methods and the more competitive organic and other sustainable agricultural methods become. A predictable long-term stable floor on the price of oil could have the additional benefit of reducing carbon emissions and stimulating research and investment in renewable energy technologies.

On the other hand, a perverse effect of higher oil prices since the early years after 2000 is stimulation of large-scale agro-industrial production of liquid biofuels for transport. These are generally produced in an agro-industrial manner with the associated negative environmental and social impacts. So the impact of higher oil prices is mixed. Putting a price on carbon that would incorporate the whole life cycle of products, including land-use changes such as deforestation to produce biofuels, would lead to better societal results. Implementation of life-cycle measurement is critically important, albeit often complex and difficult in practice.

Governments can also level the playing field between different agricultural production systems through equality in labelling requirements. The current system is perverse. The entire burden of proof and associated costs are placed on the shoulders of those farming in a sustainable, eco-friendly manner. These farmers must go through the difficulties and costs of getting certified in order to communicate with consumers that their products are produced in an environmentally sound manner. Why not instead implement labelling laws requiring declaration of all chemicals used in production and processing of the product that cause health or environmental damage or indicating when meat comes from a factory farm, and declaring sustainable production as the default?

Spectrum of support

The level of government awareness of, and engagement in, supporting sustainable agriculture including organic agriculture can be visualized as points along a spectrum as contained in Figure 8.3. Most governments start at the far left-hand side with mainly misconceptions about what organic agriculture is and can offer. Organic agriculture is rejected as old fashioned, low yielding, or non-scientific. As decision-makers in governments ingest more information and increase their levels of awareness and understanding of organic agriculture, the attitudes begin to change.

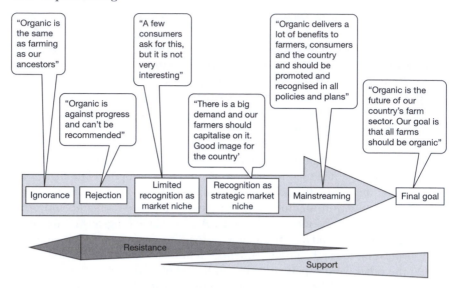

Figure 8.3 Spectrum of government awareness of and support for organic agriculture
Source: Gunnar Rundren, unpublished.

One entry point (especially in developing countries) is the recognition that the markets for certified organic products are growing many times faster than markets for conventional products and organic farmers have higher incomes. Ministries of trade usually understand this fairly quickly. Another entry point is food security. Contrary to popular myth, organic agriculture is compatible with food security (UNEP-UNCTAD, 2008b; FAO, 2002, 2007; UNHRC, 2010; 2011; Badgley et al., 2007, see also Pretty et al., 2006). Both ministries of trade and of agriculture and rural development can be interested by the improved incomes, poverty alleviation and food security contribution of organic agri-culture. Ministries of environment, though, often seem to view agriculture as beyond the scope of their influence, one reason clearly being that ministries for agriculture in many countries tend to be strong bodies with a large lobbying power among their stakeholders. When ministries of environment do get inter-ested in agriculture in general, they clearly see the negative environmental impacts of the agro-industrial model and the benefits of ecologically friendly agricultural production. Sometimes ministries of health can become interested, for example because synthetic pesticides are a top cause of occupational mortality worldwide and many agro-chemicals are carcinogenic. In Malaysia, for example, the domestic organic market initially developed largely to provide healthy food for cancer victims (Ong, 2006). But in other cases the ministry of health can work against organic farming, for example, in Uganda where it required DDT spraying to fight malaria, which contaminated organic farms and had a mixed impact on the health of the affected rural populations (Bouagnimbeck, 2010; UNEP-UNCTAD, 2011).

In general governments are not monoliths. One part of the government may be pro-sustainable agriculture, while another part may block it completely. Furthermore, different individuals in the same governmental body are at different points along the spectrum of attitudes. An individual may attend a conference, read a paper or visit an organic farm and may become supportive of organic agriculture, while their colleagues might be sceptical. In the face of the general conventional mindset of colleagues, the individual can lose their inspiration. These individuals need to be supported to keep their interest alive. They must have a chance to regularly interact with others like themselves in other governments or other parts of the government. They need information they can use to convince their colleagues. The UNEP-UNCTAD CBTF publication on organic agriculture and food security in Africa (UNEP-UNCTAD, 2008b), for example, was produced at the request of decision-makers in African governments who said they needed solid evidence on food security to overcome resistance of their colleagues, most of whom had bought into the 'organic agriculture cannot feed the world' mantra.

However, some governments in the world increasingly realize that sustainable agriculture offers a valuable contribution to the future of agriculture. These governments are moving in the direction of encouraging sustainability in their farming systems, for example, by tilting their support towards sustainable agriculture. Most European countries have some support schemes for organic farming, and also some developing countries (Gonzalvez et al., 2011). The government of South Korea, for example, gives direct payments to farmers practising environmentally friendly agriculture (EFA) with the highest payments going to organic producers. This has given a significant boost to the development of the sector: EFA production increased from 0.1 per cent of total production in 1999 to 12 per cent in 2010 (Kim, 2011). Tunisia has subsidized organic certification costs for many years with impressive results – an increase of organic production from 9,077 tons in 2002 to 246,688 in 2009 (Maamer Belkliria, 2011).

Developing supportive policies

At the request of several governments, UNCTAD and UNEP joined forces to produce the study 'Best Practices for Organic Policy: What Developing Country Governments Can Do to Promote the Organic Agriculture Sector' (UNEP-UNCTAD, 2008a, translated into French and Spanish). Based on seven case studies and other information, the study makes 35 practical and implementable recommendations covering the following areas: general policy, education and training, research, markets, standards and regulation, production and other. This section draws on the findings of that study.

Policy basics

Supporting sustainable agriculture needs to become a task of high priority within governments. Developing a policy framework that supports the development of sustainable agriculture including organic agriculture requires first and foremost political will. The political will comes from increasing the awareness of policymakers of the benefits of a shift towards sustainable agriculture and interactions with members of the organic sector. Conferences, workshops, public–private sector consultations, one-on-one interactions and short policy briefs play important roles in shifting thinking. Behind these, solid data need to be readily available in an attractive format to back up the claims. Policymakers who are stalling will often use 'we need more information' as the stalling point.

Once a critical mass of policymakers is convinced, then the work can begin in earnest. In some cases, one highly motivated, effective and well-placed individual is enough to get the momentum going.

Second, the current situation needs to be assessed professionally. Data on the current ways in which agriculture is produced in the country need to be available, and the costs in terms of environment, health and imports should be estimated. It should be assessed how current policies are affecting the incentive structures in agriculture. These policies can be directly related to agriculture, but also policies for rural development, food security, trade, environment, energy, and foreign, and industrial policy affect the structure of the agricultural sector. As already emphasized, internalization of external costs and removing perverse subsidies are most important in setting incentives for an optimal agriculture for society.

Third, challenges and obstacles as experienced in the organic sector need to be assessed in close interaction with the producers, processors, traders, researchers, NGOs, community-based organizations, etc. The key is to identify which policies are impacting them negatively and what kinds of policies would support them best in overcoming their main challenges.

Fourth, decide on and clearly communicate the main objectives of government policy and action regarding sustainable agriculture. Take a decision whether to develop a separate organic policy or integrate it into the overall policy or both. Different countries have done it in different ways. The ideal is probably to do both. The policy needs to be implemented with a clear action plan and backed with a sufficient budget. Targets, accountabilities and resources should be clearly formulated and assigned. Assigning a lead agency within the government and focal points in the other departments is also supportive for successful implementation.

To facilitate implementation of these policies, governments should also help organic farmers in accessing different types of government support that may be available at local, regional or national level. For example, farmer groups in Uganda can request extension service support on organic agriculture, but to take advantage of this opportunity they have to get organized as a group, register, fill out forms and make a good case in writing. A bit of support here could empower them a lot. Governments should support farmers in accessing Governmental services so that these services become accessible to the more vulnerable groups, including smaller-scale and female farmers as well as landless labourers, who also could benefit from training to enhance their skills and knowledge.

In the whole process of policy setting, regular communication with the organic sector is crucial. A permanent National Organic Committee (or National Sustainable Agriculture Taskforce) should be set up that brings together on a regular basis the focal points from the different government departments and organic sector stakeholders. The whole process needs to recognize the diversity in the organic sector and to ensure that there is a channel for the voices of the different types of stakeholders (for example, smallholder farmers, large-scale farmers, processors, traders, researchers, NGOs) to be heard.

Building human capital: knowledge, education and training

Supporting organic production is the first and most important task of governments. Sustainable agriculture is knowledge- and skills-intensive while less dependent on purchased external inputs. Thus, simply not using external inputs is not enough and will not work. A good organic farm involves active agro-ecosystem management. It embeds a lot of knowledge about the different plants and animals and their interrelationships. This includes which crops combine well to form a complete crop rotation or intercropping system for optimal soil fertility and pest management. For example, maize yields in Africa can be doubled by planting stem borer moth–repelling Desmodium between the maize and moth–attracting Napier grass around the field. This push–pull system has the added advantage of increased fodder for cattle and nitrogen fixation in soil (UNEP-UNCTAD, 2008b). A wealth of such knowledge can be found for Africa for example in the *African Organic Agriculture Training Manual* recently launched by Organic Africa and FiBL (Organic Africa, 2011).

Extension services and farmer field schools play a key role in transmitting this knowledge to farmers. In most countries, these services have not provided much support to sustainable/ecological/organic agriculture production techniques, although the situation is improving gradually. One successful example of information provision is the Tigray region of Ethiopia. This was one of the most environmentally degraded areas in the country. The local government teamed up with an active NGO, the Institute for Sustainable Development in Addis Ababa, to train farmers in ecological farming techniques, particularly compost making and water harvesting/soil erosion control measures. The result was a doubling of yields at the same time as purchase of synthetic agro-chemicals was reduced by 90 per cent. Moreover there was more crop matter left over for fodder for the animals (Edwards et al., 2010). Facilitating farmer–farmer exchanges is another excellent means of transmitting knowledge about sustainable agricultural production techniques.

In such initiatives, it is very important that gender aspects are duly taken into account when designing and delivering training and exchanges. In many countries women are doing most of the agricultural work and yet benefit from only a small fraction of the training, exchanges and workshops (FAO, 2011). Also female farmers are often more at ease with female extension workers. Therefore ministries of agriculture should ensure that they reach the optimal target groups and that there is gender balance in their extension staff.

Organic agriculture should be integrated into school curricula at all levels. Even primary school children can learn the basics about growing organic food. This skill will serve them well their entire lives, and an organic school garden can provide healthy inexpensive produce for school menus.[4] Organic agriculture can also be a part of government occupational retraining programmes. For example in Rwanda ex-combatants and in Kenya inmates at a prison were trained in organic agriculture production to facilitate their fruitful reintegration into society.

It is also critically important that organic agriculture be integrated into the standard curricula of agricultural universities and technical schools. Future ministry of agriculture policymakers, researchers and extension workers, as well as students who go into private sector or civil society activities, should all be fully conversant with organic agriculture production methods. Courses should be offered for existing workers to upgrade their skills. For example, government extension services should train all their extension workers in organic production techniques.

Research and innovation

Research is another key area to support (cf. Chapter 11). Only the smallest fraction of global agricultural research funds goes to organic and other sustainable agriculture production. Research is key to continually build the knowledge. It should be holistic in nature, looking at entire farm systems and the interactions between different species in time and space. It should build on traditional knowledge, practices and varieties of plants and animals and bring this rich body of knowledge together with the most recent scientific findings and modern practices. It is at that interface that the most exciting innovations occur (Twarog, 2004). It should be participatory, involving farmers and other operators as full partners in identifying areas of real need and in carrying out research.

Research should also look at appropriate technologies that can enhance production and reduce post-harvest losses. Those amount to 30–40 per cent of total produce in the South (Godfray et al., 2010). Big investments are not necessarily needed for this. For example, an organic farmer in Kenya uses an innovative system for storing maize. The maize is put into a specially built aluminium canister. Then through the burning of a candle, the oxygen in the canister is consumed thus killing off the pests. He can store this maize now for five years. He reduced post-harvest losses from 50 per cent to nearly zero with simple and sustainable technology and his family and animals enjoy food security year-round (Melonyi, 2011).

Farmers and other organic operators are also researchers and innovators. Governments should find ways to support their innovative efforts and also to share the outcome of their innovations with others. This can include technologies, seeds, breeds, techniques as well as innovative ways of working cooperatively and marketing their products. It can also include insurance against failure, as innovation is an inherently risky endeavour. Organic operators can be connected to innovation incubator programmes in other parts of the government or to private sector actors that can help them to scale up and disseminate their idea or product.

Market development

Governments have an important role to play in the development of local, national, regional and international markets for organic products. A key role is to raise awareness among both producers and consumers on the benefits of organic agriculture production and products. Policymakers can talk about organic agriculture in their speeches and write about it in their documents. This is a cost-effective way to stimulate both supply and demand sides of the market. Governments can also procure organic products for government institutions (offices, schools, hospitals, etc.) and feature organic products at government events. This can have a big impact in light of the importance of government procurement in most countries: it accounts for over 17 per cent of GDP in 19 OECD countries and up to 40 per cent of GDP and 70 per cent of government spending in some developing countries (OECD, 2009; Ellmers, 2010).

Governments can support the development of the organic supply chain in dialogue and cooperation with sector stakeholders, including farmers (cf. Chapters 6 and 7). For example, they can help farmers' groups to get organized to coordinate production schedules, coordinate transport of surplus products from their farms to market, jointly market their products, meet potential buyers, get the training they need, and add value to their products, for example, through joint purchase and operation of solar dryers or storage facilities. Being able to process or properly store products increases farmer incomes because they do not have to sell at the moment when the harvests are coming in on all the farms in the region and prices are low. They can wait until prices are higher (for further discussion, see Twarog, 2008).

There is thus not only a need to develop knowledge and expertise in organic production, but also a need to develop the capacities of organic entrepreneurs. Worldwide organic markets are expanding in diverse ways and opportunities – and also risks – abound. All over the world, there is a need for skilled and innovative people who can bridge the gap between organic production and connections to the market and market development. Existing entrepreneurs should have access to training on organic production and markets and interested organic operators should have access to training to become more entre-preneurial.

Standards and market regulation

A big question is the focus of government action. Many governments make the mistake of starting off focusing on regulating the use of the word 'organic'. Often this is in response to a few vocal voices from the organic sector itself, who do not always think through what would be the best possible use of limited government resources and effort that could be dedicated to the organic sector.[5]

Such regulation should only be a second stage, after the organic sector has reached a certain level of development. In the initial phases, it is most important to support the development of organic production, processing and marketing. Premature regulation risks strangling the sector before it even gets a chance to get out of the birthing bed.

If the sector has grown to a first, stable basis for further development, a definition of what organic agriculture means at national level can be useful. This can be in the form of a national or regional organic agriculture standard. Such a standard should be developed in an open consultative way in partnership with the organic sector in the country or region. This ensures that the standard is well adapted to local conditions and aspirations, and also builds a sense of ownership and momentum for the sector. The standard should also take into account what has been done in already existing standards and international guidelines (see below). The common standard helps everyone to 'be on the same page'. It can be used for training materials and sales on the domestic or regional market as it serves as a quality benchmark for organic products.

For example, the East African Organic Products Standard (EAOPS) was developed between 2005 and 2007 through a highly consultative regional public–private sector partnership, facilitated by the UNEP-UNCTAD CBTF (Capacity Building Task Force) and IFOAM. Members of the technical working group from public and private sectors of the five East African countries (Burundi, Kenya, Rwanda, Uganda and the United Republic of Tanzania) met four times to harmonize the requirements in the six standards already existing in the region. Drafts of the EAOPS were widely consulted in six national multi-stakeholder consultations as well as international calls for comments. The resulting standard is in line with international organic guidelines, and at the same time reflects uniquely East African perspectives and conditions. For example, it requires that wildlife corridors be provided for migratory species and that no primary ecosystem be cleared for organic farms. The EAOPS was adopted by the East African Council of Ministers in 2007 and launched, along with the associated East African Organic Mark, by the Prime Minister of Tanzania (East African Community, 2007; UNEP-UNCTAD, 2010, 2011; IFOAM, 2008; OECD-WTO, 2011).

The EAOPS process served as an inspiration for other regional harmonization efforts. This includes the development of the Pacific Organic Standard finalized in 2008 (Secretariat of the Pacific Community, 2008) and the current process to draft the Asian Regional Organic Standard which started in 2010 with the facilitative support of the UNCTAD-FAO-IFOAM Global Organic Market Access initiative (GOMA, 2011).

A standard can be adopted by the national government as the official organic standard of the country. It is not necessary at the same time to immediately adopt a heavy conformity enforcement regime requiring third-party certification by accredited certification bodies. For local markets, trust between producer and consumer has been the basis for sales for millennia. Options for guaranteeing the organic claim should be left open. These options include:

- self-claim, especially for smaller operations; in the United States, for example, organic regulation does not require third-party certification for direct farmer sales of less than US$5,000;
- second-party peer-to-peer verification systems such as participatory guarantee systems (PGS) (IFOAM, 2011);
- third-party certification for more formal and more distant markets, including those in other countries; and
- group certification where groups of smaller operators self-regulate through internal control systems, and then get certified by an external certifier, who ensures that the internal controls are functioning properly, including an inspection of a certain proportion of operators.

International level

Both the public and private sector should participate in international standard setting in organic agriculture and related fields. This includes keeping an eye on the international standards as well as standards, regulations and conformity assessment requirements in current or potential trade partners. Often international standards are opened for comment and only a few people or institutions respond.

Governments should ensure that their organic regulations are open to trade – both imports and exports. Imports play a very valuable role in the development of the domestic organic sector. Consumers become more interested in purchasing organic products (becoming organic consumers) when they have an array of products available to choose from. Moreover, ingredients for processed products often must be sourced from many different countries.

Equivalency should be the basis of international trade in organic products, supplemented by harmonization where desired and applicable. National organic standards and regulations should be in line with international organic standards and also, very importantly, be tailored to national agro-ecological, socio-economic and cultural perspectives. International trade should be based on mutual respect for this policy space. Countries and private sector standard setters should not force the rest of the world to comply in a prescriptive manner with their organic guarantee systems, which might not fit the other countries' context. The way forward is to expect the best while at the same time embracing diversity. Countries should allow imports of organic products that are produced and guaranteed in a manner equivalent, but not necessarily identical, to their own.

UNCTAD, FAO and IFOAM have worked together and with a host of key public and private sector actors for over a decade to develop tools to make trade based on equivalency happen. For conformity assessment, the International Requirements for Organic Certification Bodies (IROCB), which are performance requirement objectives for organic certifiers adapted from ISO 65, facilitate recognition across systems. For production and processing standards, the EquiTool is a guide to assessing differing standards in a structured and transparent manner. Both these tools were developed by the FAO-IFOAM-UNCTAD International Task Force on Harmonization and Equivalence in Organic Agriculture (ITF) (2002–2008).[6]

Under the successor joint initiative, the Global Organic Market Access (GOMA) project, the EquiTool has been enhanced through the development of the Common Objectives and Requirements for Organic Systems (COROS). COROS helps governments and other organic standard setters to identify the underlying objectives their systems are aiming to achieve and then to evaluate other standards to see if, on the whole, they achieve those objectives (in a similar or different but equally valid manner). IFOAM, the international private sector standard setting body, is using COROS to develop the IFOAM family of standards – those which have been assessed and found to be overall equivalent to COROS. For more information and to download COROS see the GOMA website (UNCTAD-FAO-IFOAM, 2011).

Therefore, for the purpose of trade in organic products, and particularly regarding imports of organic products from other systems, public and private sector actors involved in regulating organic guarantee systems should:

For production and processing standards
• Use COROS and the Equitool to evaluate the other production and processing standards to determine if compliance with it would, as a whole, achieve the most important underlying objectives of organic production systems.

For conformity assessment systems
• Build trust among accreditors and supervising bodies (including governments) to mutually recognize accreditation/approval systems of certification bodies and other means of conformity assessment.
• Use IROCB to evaluate the certification bodies' performance requirements.

Many regions are undergoing harmonization of their organic production and processing standards. The European Union (2007), East African Community (2007) and the Pacific Islands (SPC, 2008) have developed and adopted regional standards. With support of the GOMA project, the Central American countries plus the Dominican Republic and countries from South, Southeast and East Asia are in the process of drafting and consulting their regional standards (UNCTAD-FAO-IFOAM, 2011).

Import regulations should allow for recognition of regional standards. Currently this is not the case. For example, under the EU system there is no way for the East African Community to submit the EAOPS for approval because a common organic conformity assessment system replete with accreditors and supervision of certification has not yet been developed. So the region is not part of the so-called third countries whose organic standards are accepted by the EU. Yet this is the official organic standard of five countries. The EU and other regulations should allow for the option to recognize regional organic standards in the regulation of organic imports.

Conclusions

There are powerful reasons for governments to make strong commitments and take strong actions to support the development of organic agriculture. There are also powerful vested interests that spend many millions of dollars to convince these same governments that industrial agriculture based on the purchase of their products is the only possible way to feed the world.

I call upon policymakers to be strong, independent and wise. When someone gives you information, ask yourself what is their underlying objective or interest. If they tell you that you must buy their product or you will be morally responsible for the starvation of your population, question this! Their corporate or country interests may not be in the best interests of most citizens in your country. If they show you data showing impressive productivity gains in two years, ask what happens after ten years. And do not only ask them. Get a second and third opinion. Ask farmers and NGOs in areas that tried to follow their advice for years. Seek the advice of those who have no narrow economic self-interest in your decision.

Ask the United Nations. It is no accident that the UN Special Rapporteur on the Right to Food, as well as UNCTAD, UNEP, FAO, UN Department for Social and Economic Affairs, International Trade Centre and many others have signaled that sustainable ecological agriculture including organic agriculture is the most promising path for the future.

Notes

1   Bayer Crop Science, Syngenta, BASF, Monsanto, Dow Agrosciences and Dupont/Pioneer.
2   This has long been recognized in economics. See the work of Mancur Olson Jr. (1965) for example.
3   For example, Sub-Saharan Africa imports 99 per cent of mineral fertilizer consumed – 1,031,591 out of 1,040,992 metric tons in 2005/2006 (IFDC, 2008).
4   The Soil Association has pioneered such initiatives and provides a full set of materials on how to do this at www.foodforlife.org.uk.
5   Later, these same people sometimes regret that they got the government so involved in policing the organic sector and rather wish they had been able to work out their problems among themselves.
6   An analysis of Asian systems revealed that the IROCB more closely described what was actually required by governments in assessing performance of organic certification. IROCB is a sector-specific adaptation of ISO 65. It leaves out a few items considered irrelevant for organic certification and adds some items deemed necessary for a proper guarantee of organic integrity. It was developed through a highly consultative process. Both IROCB and Equitool are available on the ITF website at www.unctad.org/trade_env/itf-organic and the GOMA website at www.goma-itf.org.

Disclaimer

The views expressed in this chapter are those of the author and do not necessarily reflect the views of UNCTAD.

200   *Sophia Twarog*

References

Agriculture and Agri-Food Canada (2010) 'Canadian farm fuel and fertilizer: prices and expenses', *Market Outlook Report* 2(7), November 26, http://www.agr.gc.ca/pol/mad-dam/index_e.php?s1=pubs&s2=rmar&s3=php&page=rmar_02_07_2010-11-26

Badgley, C., Moghtader, J., Quintero, E., Zakem, E., Jahi Chappell, M., Aviles-Vazquez, K. et al. (2007) 'Organic agriculture and the global food supply', *Renewable Agriculture and Food Systems*, 22, 2, pp. 86–108. For a criticism of this study and response from the authors see the same journal, 22, 4, pp. 321–329

Bellarby, J., Foereid, B., Hastings, A., Smith, P. (2008) 'Cool farming: climate impacts of agriculture and mitigation potential', Greenpeace International, Amsterdam, the Netherlands. www.greenpeace.org/international/Global/international/planet-2/report/2008/1/cool-farming-full-report.pdf

Bouagnimbeck, H. (2010) 'Organic farming in Africa', in H. Willer and L. Kilcher (eds), *The World of Organic Agriculture: Statistics and Emerging Trends 2010*, IFOAM and FiBL, Bonn and Frick

Center for Food Safety (2005) 'Monsanto vs. U.S. Farmers', www.centerforfoodsafety.org/pubs/CFSMOnsantovsFarmerReport1.13.05.pdf

Chemtox (2011) 'Chemical pesticides health effects research', Website providing summaries of many peer-reviewed studies on this topic, www.chem-tox.com/pesticides

Convention on Biological Diversity (CBD) (2010) Decision adopted by the Conference of the Parties to the Convention on Biological Diversity at its tenth meeting. X/2. The Strategic Plan for Biodiversity 2011–2020 and the Aichi Biodiversity Targets. UNEP/CBD/COP/DEC/X/2. 29 October.

East African Community (EAC) (2007) 'East African organic products standard', EAS 456:2007

Easterling, W.E., Aggarwal, P.K., Batima, P., Brander, K.M., Erda, L., Howden, S.M., et al. (2007) 'Food, fibre and forest products'. in M.L. Parry, O.F. Canziani, J.P. Palutikof, P.J. van der Linden and C.E. Hanson, (eds.), *Climate Change 2007: Impacts, Adaptation and Vulnerability*. Contribution of Working Group II to the Fourth Assessment Report of the Intergovernmental Panel on Climate Change, Cambridge University Press, Cambridge, UK, 273–313

Edwards, S., Egziabher, T. and Araya, H. (2010) 'Successes and challenges in ecological agriculture, experiences from Tigray, Ethiopia'. FAO, Rome

Ellmers, B. (2010) 'Eurodad update: procurement, tied aid and the use of country systems', 1 April, European Network on Debt and Development, www.eurodad.org/whatsnew/articles.aspx?id=4058.

Elser, J. and White, S. (2010) 'Peak phosphorus', *Foreign Policy*, 20 April, www.foreignpolicy.org

European Union (2007) 'Council Regulation (EC) No 834/2007 of 28 June 2007 on organic production and labelling of organic products and repealing Regulation (EEC) No 2092/91' *Official Journal of the European Union*, 189/1, pp. 1–23

FAO (2002) 'Organic agriculture, environment and food security', FAO, Rome

FAO (2007) 'Report of the International Conference on Organic Agriculture and Food Security', Rome, 3–5 May, OFS/2007/REP, ftp://ftp.fao.org/docrep/fao/meeting/012/J9918E.pdf

FAO (2011) 'The State of Food and Agriculture 2010–11, Women in Agriculture: Closing the gender gap for development'. FAO, Rome

Godfray, H., Beddington, J., Crute, I., Haddad, L., Lawrence, D., Muir, J., et al. (2010) 'Food security: the challenge of feeding 9 billion people', *Science,* 327, 5967, pp. 812–818

GOMA (2011) For more information, see www.goma-organic.org

Gonzalvez, V., Schmid, O. and Willer, H. (2011) 'Organic action plans in Europe in 2010', in H. Willer and L. Kilcher (eds), *The World of Organic Agriculture: Statistics and Emerging Trends 2010*, IFOAM and FiBL, Bonn and Frick

Hoffmann, U. (2011) 'Assuring food security in developing countries under the challenges of climate change: key trade and development issues of a fundamental transformation of agriculture', UNCTAD Discussion Paper No. 201 February. HRC (2010) Report submitted by the Special Rapporteur on the Right to Food, Olivier de Schutter, A/HRC/16/49, UNCTAD/OSG/DP/2011/1

IFOAM (2008) 'Development of a regional organic agriculture standard in East Africa 2005–2007', www.ifoam.org/partners/projects/osea/pdf/OSEA_Project_Report.pdf

IFOAM (2011) 'Participatory Guarantee Systems', International Federation of Organic Agriculture Movements, www.ifoam.org/about_ifoam/standards/pgs.html

International Fertilizer Development Center (IFDC) (2008) 'Fertilizer statistics', www.ifdc.org/Media_Info/Fertilizer_Statistics

Jimenez, J. (2006) 'Organic agriculture and the Millenium Development Goals', International Federation of Organic Agriculture Movements (IFOAM), Bonn

Kilcher, L. and Maamer Belkhiria, S. (2011) 'Tunisia: country report', in H. Willer and L. Kilcher (eds), *The World of Organic Agriculture: Statistics and Emerging Trends 2010*, IFOAM and FiBL, Bonn and Frick

Kim, S.H. (2011) 'Korean organic agriculture: models for a healthier planet', Key Note Address at Organic World Congress (Seoul, 28 September)

Matson, P.A., Parton, W.J., Power, A.G. and Swift, M.J. (1997) 'Agricultural intensification and ecosystem properties', *Science*, 277, 5325, pp. 504–509

Melonyi, P. (2011) Personal on-site interview, Kenya

Monsanto (2011) Percy Schmeiser, www.monsanto.com/newsviews/Pages/percy-schmeiser.aspx

Myers, D. (2000) 'Cotton Tales', *New Internationalist* 323, www.newint.org/issue323/tales.htm

OECD (2009) *Government at a glance 2009*, OECD Publishing, Paris

OECD-WTO (2011) 'Aid for trade case story: the East African organic products standard', Submission by the United Nations Environment Programme, www.oecd.org/dataoecd/11/62/47719232.pdf

Olson, M. (1965) *The Logic of Collective Action: Public Goods and the Theory of Groups*, Harvard University Press, Cambridge, MA

Ong, K.W. (2006) 'Opportunities and woes: public–private partnership in development of the organic sector in Malaysia', Paper presented at UNCTAD/ UNESCAP workshop on Maximizing the Contribution of Organic Agriculture to Achieving the Millennium Development Goals in the Asia and Pacific Region (Bangkok), www.unctad.org/trade_env/meeting.asp?MeetingID=178

Organic Africa (2011) *African Organic Agriculture Training Manual*, www.organic-africa.net/training-manual.html

Pimentel, D. (1997) 'Livestock production: energy inputs and the environment', in Scott, S.L., and Zhao, X. (eds) *Canadian Society of Animal Science, proceedings*, Vol 47. Montreal, Canada pp. 17–26

Pretty, J.N., Noble, A.D., Bossio, D., Dixon, J., Hine, R.E., Penning de Vries, F.W.T. and Morison, J.I.L. (2006) 'Resource-conserving agriculture increases yields in developing countries', *Environmental Science and Technology*, 40, 4, pp. 1114–1119

Sahota, A. (2011) 'The global market for organic food and drink', in H. Willer and L. Kilcher (eds), *The World of Organic Agriculture: Statistics and Emerging Trends 2011*, Frick: FiBL; Bonn: IFOAM

Sawyer, L. (2003) 'Natural gas prices impact nitrogen fertilizer costs', *Integrated Crop Management*, 490, 4, pp. 32–33

Schmeiser, P. (2011) 'Monsanto vs. Percy Schmeiser', www.percyschmeiser.com/conflict.htm

Secretariat of the Pacific Community (SPC) (2008) *Pacific Organic Standard*, Noumea, New Caledonia

Setboonsarng, S. (2006) 'Organic agriculture, poverty reduction and the Millenium Development Goals', ABD Institute Discussion Paper No. 54

Smith, P., Martino, D., Cai, Z., Gwary, D., Janzen, H., Kumar, P., et al. (2007) 'Agriculture', in B. Metz, O.R. Davidson, P.R. Bosch, R. Dav and L.A. Meyer (eds), *Climate Change 2007: Mitigation of Climate Change*. Contribution of Working Group III to the Fourth Assessment Report of the Intergovernmental Panel on Climate Change, Cambridge University Press, Cambridge, UK

Smith, P., Martino, D., Cai, Z., Gwary, D., Janzen, H., Kumar, P., et al. (2008) 'Greenhouse gas mitigation in agriculture', *Philosophical Transactions of the Royal Society B*, 363, pp. 789–813

Soil Association (2010) 'A rock and a hard place: peak phosphorus and the threat to our food security', www.soilassociation.org

Soussana, J.F., Tallec, T. and Blanfort, V. (2010) 'Mitigating the greenhouse gas balance of ruminant production systems through carbon sequestration in grasslands', *Animal* 4, 03, pp. 334–350

Steinfeld, H., Gerber, P., Wassenaar, T., Castel, V., Rosales, M. and C. de Haan (2006) *Livestock's long shadow*, Food and Agriculture Organization of the United Nations, Rome

Sutton, M.A., Howard, C.M., Erisman, J.W., Billen, G., Bleeker, A., Grennfelt, P., et al. (eds) (2011) *The European Nitrogen Assessment: Sources, Effects and Policy Perspectives*, Cambridge University Press, Cambridge, UK

Tilman, D., Cassman, K., Matson, P., Naylor, R. and Polasky, S. (2002) 'Agricultural sustainability and intensive production practices', *Nature*, 418, pp. 671–677

Turner, K., Georgiou, S., Clark, R., Brouwer, R. and Burke, J. (2004) 'Economic valuation of water resources in agriculture', FAO paper reports No. 27, FAO, Rome

Twarog, S. (2004) 'Preserving, protecting and promoting traditional knowledge: national actions and international dimensions', in S. Twarog and P. Kapoor (eds), *Protecting and Promoting Traditional Knowledge: Systems, National Experiences and International Dimensions* (UNCTAD/DITC/TED/10), United Nations, New York and Geneva

Twarog, S. (2006) 'Organic agriculture: a trade and sustainable development opportunity for developing countries', in UNCTAD, *Trade and Environment Review 2006*, UN, New York and Geneva, www.unctad.org/en/docs/ditcted200512_en.pdf

Twarog, S. (2008) 'Trade and sustainable development options for smallholder farmers in light of global trends: the case of organic agriculture', www.unctad.org/trade_env

Twarog, S. and Kapoor, P. (eds) (2004) *Protecting and Promoting Traditional Knowledge:*

Systems, National Experiences and International Dimensions (UNCTAD/DITC/TED/10), United Nations, New York and Geneva

UNCTAD-FAO-IFOAM (2008) Website of the UNCTAD-FAO-IFOAM International Task Force on Harmonization and Equivalence in Organic Agriculture (ITF), www.unctad.org/trade_env/itf-organic

UNCTAD-FAO-IFOAM (2011) Website of the Global Organic Market Access (GOMA) project, www.goma-itf.org

UNEP (2011) 'Agriculture: investing in natural capital', Green Economy Report

UNEP-UNCTAD (2008a) 'Best practices for organic policy: what developing country governments can do to promote the organic agriculture sector' (UNCTAD/DITC/TED/2007/3). United Nations, New York and Geneva. www.unctad.org/trade_env

UNEP-UNCTAD (2008b) 'Organic agriculture and food security in Africa' UNCTAD/DITC/TED/2007/15). United Nations, New York and Geneva

UNEP-UCTAD (2010) 'Organic agriculture: opportunities for promoting trade, protecting the environment and reducing poverty: case studies from East Africa', Synthesis of the UNEP-UNCTAD CBTF Initiative on Promoting Production and Trading Opportunities for Organic Agriculture in East Africa (DTI/1225/GE), www.unep.ch/etb/publications/Organic%20Agriculture/OA%20Synthesis%20v2.pdf

UNHR (2011) News Release, 'Eco-Farming can double food production in 10 years, says new UN report', and Human Rights Council (2010). Report submitted by the Special Rapporteur on the Right to Food, Olivier de Schutter. (A/HRC/16/49)

USDA (2011) 'Economic Research Service Briefing Room: global food markets', www.ers.usda.gov/Briefing/GlobalFoodMarkets

US Geological Survey (2010) 'Mineral commodity summaries', January

White, S. and Cordell, D. (2010) 'Peak phosphorous: the sequel to peak oil', Available from phosphorousfutures.net (full article in *Global Environmental Change* journal)

World Bank (2011) World Bank Commodity Price Data, http://data.worldbank.org/data-catalog/commodity-price-data

Zengxin (2008) 'China announces export tax, tariff adjustments', *China Daily* 21 February

9 Comparative institutional analyses of certified organic agriculture conditions in Brazil and China

Henrik Egelyng, Lucimar Santiago de Abreu,
Luping Li and Maria Fernanda Fonseca

Introduction

Studies of institutional factors influencing organic farming in the South have gained in importance in response to the ongoing globalization of the organic food system. Drivers and policy initiatives differ across countries and regions and the degrees and ways in which organic principles are locally embedded varies. This chapter analyses the general conditions for Certified Organic Agriculture (COA) in Brazil and China, exploring the history and scope of their national systems of regulation of and support for COA. The levels and nature of public and private agency are determined through the use of qualitative indicators covering a range of institutions, from public policy instruments such as national action plans, laws and public research through to private and civil society initiatives. The goal of this chapter is to present a framework for analysis of national institutional environments for development of organic sectors and findings from case studies of Brazil and China based on the same framework. Finally, analysis is made of differences and similarities between the institutional pathways of the organic sectors in the nations compared, identifying social and economic policy conditions under which certified organic production evolves at national and sub-national levels ('glocalization') in the context of globalization. Hosting the Olympic Games is one of the factors of shared importance. The chapter concludes that while both countries have an advanced set of institutions to serve certified organic agriculture, COA is quite differently embedded in Brazil and China and so are policy options.

Analytical framework

Our analytical and soft comparative framework has the following five dimensions: (I) overall policies, (II) regulation – in particular conformity assessment systems; (III) research, education and extension that targets COA; (IV) agency and the roles of the private sector and civil society organizations (CSOs); and

(V) a broader contextual analysis. The first dimension focuses on the nature of overall national policy concerning COA: law(s) on organic agriculture and the extent to which strategies are translated into national action plans on COA. The second dimension assesses the institutional localization of responsibilities for the development of organic standards, certification and accreditation, and other aspects of conformity assessment recognized in organic regulations. The third dimension investigates the extent and nature of organic research and education at agricultural universities and colleges, and the extension provided by public and private organizations. The fourth dimension determines whether, to what extent and how the private sector, including companies, farmers' organizations and other CSOs, undertakes activities and assumes a de-facto policy development role towards organic agriculture. Finally, the fifth dimension provides data on the contextual environment for COA and explores how policy goals on organic agriculture sit within the overall agrarian and rural development strategy. This framework aims to provide indicators as to whether organic farming finds itself in a 'policy ghetto' or is more or less integrated in other policy areas such as tax, environment, rural development, and health and consumer policy. It particularly focuses on the balance between command and control rules and regulation, economic instruments and information. The extent to which the polluter pays principle influences agriculture, so as to possibly help level the playing field for COA, is also included in our framework for exploration.

Comparative studies can rely on different strategies, such as comparing most similar or most different cases. To avoid any methodological straightjacket, we adopted a soft comparative approach and the above framework was applied to structure literature studies, data collection and exploratory interviews undertaken in the course of 2006 and completed in January 2007. Below we present our findings in the form of broad explanations on the variations in institutional pathways between the two nations.

Brazil

Already emerging as one of the giants of the world of organic agriculture, Brazil is expected to have at least doubled its production and sales of organic products, once the full impact of hosting the 2014 FIFA World Cup and the Olympic Games in 2016 is felt. Detailed data about COA are dispersed among civil society and certification bodies (CBs) and, with the exception of a few states such as Paraná, there has been no systematic government register. Early estimates of the area under organic management and the number of certified or compliance-assessed organic farms have therefore continued to be a matter of debate and conflicting estimates. A new information system established by the Ministry of Agriculture, Livestock and Food Supply (MAPA) is perhaps the latest and most qualified estimate: 15,000 organic farmers officially (and centrally) registered as of 2011. For more details on the agro-ecological movement, which plays an important role in Brazil, see Chapter 10.

I Overall policy

Intense consultation with stakeholders and public participation have provided Brazil with a broad and inclusive concept of organic production, established in law (No. 10,831). The Organic Agriculture Productive Chain Sector Chamber (CSAO), consisting of public and private representatives of the organic sector, is well established as an advisory body for MAPA. This chamber, along with an Inter-Ministerial Commission for Agroecology and Organic Production Systems, is the major institutional basis for laws and regulations influencing the organic sector. MAPA's Organic Agriculture Development Programme (PRO ORGÂNICO) supports production, manufacture and commercialization of COA products. This includes an action plan with roles for all of MAPA's units, states, municipal authorities and CSOs, including CSAO, the Organic Production Commissions in federal units (CPOrgs) and the Organic Production National Commission (CNPOrg).

The Ministry of Agrarian Development (MDA) supports development of COA by giving financial support for conversion from conventional agriculture for a financially and legally well-defined category of 'family farmers'. These are farmers, who generate at least 80 per cent of their household revenue from agriculture, have a maximum of four area units (the size of which differs across regions, districts and agro-ecosystems) and no more than one employee. Land redistributed under recent agrarian reforms is also eligible for special financial support for organic agriculture. The programme facilitates smallholders' participation in the process of national regulation and provides funds to support networks of organic farmers and smallholders. National policy also provides for technical advisers and rural extension services to support sustainable production systems, with specific financing for organic production via a national policy for family agriculture and rural family enterprises (PRONAF Agroecologia). The MDA has partnerships with the German Technical Cooperation agency (GTZ) and the Slow Food movement. Environment Ministry (MMA) programmes for natural resource conservation also support organic agriculture. The Brazilian accreditation institute (INMETRO) and the National Agency for the Development of Small and Medium Enterprises (SEBRAE) help organic producers meet certification costs and the Trade and Investment Promotion Agency (APEX) promotes Brazilian organic products at relevant international trade shows. Finally, the Bank of Brazil provided financial lines to COA initiatives in regions with threatened agro-ecosystems.

The levels of awareness about the labels, principles and benefits of COA were moderate to low during the first decade of the new millennium (Guivant et al., 2003; Darolt, 2004). Today, the certified organic agricultural sector in Brazil is benefiting from the planning of the two major global sports events that Brazil hosts: the 2014 FIFA World Cup and the 2016 Olympic Games. Twelve Brazilian states are involved in building or strengthening the capacity of organic producers in their jurisdiction to be able to supply enough certified organic food to meet the demand from all the expected visitors from countries where consumption of organic products are well established. The nature of the challenge can be illustrated by the fact that in one state – Minas Gerais – initially only 641 out of 12,000 expected food producers/suppliers proved to be formally certified or conformity assessed. Therefore, agricultural authorities grasped the opportunity to intensify and improve the flow of information on COA and the organic sector law (No. 10,831) and organized a cycle of courses and lectures in the form of a so-called 'Caravan Cup Organic Brazil 2014', sponsored by SEBRAE as well as by private companies. The 'Cup' is touring 12 states and 44 major cities – like Belo Horizonte, Jaboticatubas and Betim in Minas Gerais (caravancopaorganica.com.br). At the federal level the Ministry of Sports has included organizations and consortia of organic producers in preparation meetings. It is estimated that the two events will double the market for organic products in Brazil compared to its US$400 million level prior to them.

These developments have further strengthened national, state and municipal policies of 'buying organic', which is increasing both the organic market and awareness. One example is the Family Farmers' Food Acquisition Programme (PAA), a partnership between the Ministry of Social Development (MDS) and MAPA, implemented by the National Supply Company (CONAB). CONAB pays a 30 per cent premium for COA products. This premium is justified in order to preserve agro-biodiversity and environmentally sustainable management systems. In some rural areas government authorities promote organic agriculture for its beneficial effects on food security. Organic 'knowledge centres' (such as ABD, Fundação Mokiti Okada, Centro Ecológico, ASSESSOAR) advise the private sector and supermarkets in major cities offer specialized Internet sites for COA (e.g. www.planetaorganico.com.br). In Paraná state, an organic school meal programme funded by the Environment Secretary distributes locally produced organic products to 66 municipalities. This programme aims to reduce the use of agro-chemicals, facilitate smallholder access to new commercial chains and stimulate local food consumption. A similar programme exists in Santa Catarina State, where the Secretary of Education, in partnership with farmers' associations, established a programme for supplying local organic foods for school lunches. Set up in 2002, this programme has since then provided organic lunches for an increasing number of school children.

II Regulatory set up

Along with so-called 'normative instructions', the organic sector law facilitates use of the international IFOAM-based standards normally implied by organic certification, as well as alternative types of conformity assessment procedures. In Brazil, therefore, some producers are conformity assessed under Participatory Guarantee Systems (PGS) – stakeholder-oriented systems, aiming to be flexible vis-à-vis different social, cultural, political and economic realities, based on supplier conformity declaration and verification by peer review, as an alternative to formal certification (Fonseca et al., 2006). The normative instructions regulate specific areas such as production and marketing of organic seed and seedlings, marketing, transportation and storage of organic products and organic textile production. An office of evaluation of organic compliance (OAC) is involved in the verification of organic products, producers and service providers. The first Brazilian private organic standards, based on international organic standards, were established in the 1980s. Ten years later, when Brazil started to export organic products to Europe, pressure for the establishment of a Brazilian authority of certification body and for national legislation followed. The process for formally regulating COA at the federal level was initiated in 1994, following pressure from civil society organizations and reached Congress in 1996. There was intense debate in the organic movement about the inclusion of group certification and participatory guarantee systems when MAPA provided the first generation of 'normative instructions' regulating the production, manufacturing, labelling and certification of organic products. This included provision for PGS and group certification. Subsequent attempts to ratify these elements have continued to be the source of debate. Existing rules allow for both group certification and PGS, as three interim 'consensus' documents from MAPA are operating. The formal rules are therefore best characterized as broadly following international standards, having particularities however about conversion period, 'social justice', wild harvesting standards and criteria for conformity-assessment systems, including PGS, MAPA, INMETRO and CSOs have emerged as sharing responsibility for the Brazilian System of Organic Conformity Assessment (SISORG). The system is managed by state and national commissions. INMETRO is in charge of accrediting Certification Bodies (CBs), based on ISO-65 standards and according to Brazilian organic regulations. For family farmers following a social control process for direct sales, certification is not mandatory. Family farmers, however, have to be members of a CSO, registered with MAPA or an equivalent body authorized at the local level. For marketing purposes organic products must be produced in accordance with the Brazilian regulation and certified by an accredited CB.

Three kinds of certification bodies operate in Brazil: international CBs with or without a Brazilian office; national CBs with (or in the process of gaining) international recognition for accessing the main markets (in the USA, EU and Japan) and/or accreditation from private international organic standards (IFOAM Organic Guarantee System); and CBs that evolved from organic farmers' and advisers' associations. In addition, some national organizations work with certification and PGS, and commercialization of internal markets. As many as 30 farmers' associations and CBs (10 international and 20 national) have been involved in certification (Fonseca et al., 2006), although today (2012) this figure is estimated to be lower.

III Research policy

The involvement of (state) governments in research on organic agriculture officially began in 1988, when the State Agricultural Research Corporation (PESAGRO-RJ) established an experimental centre in Rio de Janeiro. At the national level, the Brazilian Agricultural Research Corporation (EMBRAPA) established a programme for organic agriculture in 2000. EMBRAPA operates a field station which includes organically managed agricultural research land, and this has provided an important reference point for organic and agro-ecological production systems since 1992. The National Council of State Agricultural Research Enterprises (CONSEPA) is a consortium of 17 research and development (R&D) organizations, and has at least 40 researchers and advisers who work directly in R&D for COA. There is also involvement in COA at the municipal and state level.

IV Civil society

Promotion of organic food weeks is regularly undertaken with workshops and events in major cities, often jointly sponsored by MAPA, MMA and MDA, the Brazilian supermarket association (ABRAS), and other public and private stake-holders. Information on organic agriculture is provided by NGOs, the private sector and government bodies. This is aimed at students, farmers, technicians, retailers and consumers. Programmes about food and the economic possibilities for COA are broadcast on television. Downloadable documents about organic agriculture are available at main research institutes, organic agriculture legislation is available at government and commercial sites, and national journals are published on organic agriculture including both scientific ones by public research organizations and universities and private magazines such as *Boletim Agroecológico*.

The first ecological (Saturday) street market was organized by COOLMEIA, a farmers' and consumers' cooperative, initially involving 25 farmers from different parts of the state of Rio Grande do Sul. ECOVIDA (Rede Ecovida de Agroecologia) – a network created with member organizations initially from the three southern states in Brazil, organized in 23 nodes and involving 2,600 families and 290 groups of small farmers, retailers and consumers cooperatives – has contributed very significantly to the commercial success of organic marketing in Brazil. Historically, therefore, the most valuable commercial channels were local street markets and institutional markets such as schools, which accounted for about two-thirds of total sales volume (Santos, 2004).

The early price premiums on Brazilian organic products ranged from 20 to 250 per cent depending on the product and commercial chain, although only a small share was passed on to producers (Guivant et al., 2003). The costs of external audits, implementing control systems and investment in training personnel were identified as bottlenecks for national and international accreditation (Medaets and Fonseca, 2005) along with commercialization problems such as high costs of certification, high levels of rejection (out-grades), packaging and transport logistics. Smallholder group certification was launched as one strategy for reducing transaction costs, as this decreased certification costs for each group member by up to 35 times (Medaets, 2003). The direct cost to farmers of participatory certification is significantly lower for smallholder group certification, but the indirect costs (organization, technical advice and capacity building) are higher. These costs are covered by voluntary work of farmers, technicians, sympathizers and consumers (Medaets, 2003). In a PGS, the members contribute a small monthly fee (US$6–17 per month). In exchange they receive resources from government sustainable development projects or from international agencies, to cover internal controls, meetings, visits and registers (Meirelles, 2004; Nunes, 2004; Santos, 2004).

Some MDA projects support civil society participation in discussions about how to adapt the organic regulation to Brazilian circumstances. Others help national CBs, mainly those working with smallholders producing for domestic markets, to adapt their management to ISO standards, so that they can receive accreditation. MDA has also supported organized groups of family farmers seeking group certification, providing support for developing PGS standards that will meet Brazilian criteria for Organic Conformity Assessment (SISORG) and for Fair Trade.

V Context

MMA has proposed new environment policy instruments to the Brazilian Congress. These include changing articles in national environmental and agricultural legislation (No. 6,938 and No. 8,171 respectively) and introducing tariffs to provide incentives for production activities that are environmentally sustainable. Discussions on GMO regulations and eco-taxation are ongoing between government and the CSO. There are mechanisms to link organic farming with tourism, which are being supported by MDA using resources from PRONAF. Several states use funds collected through sales tax for environmental purposes including the promotion of organic agriculture.

China

China is another giant in the new world of certified organic agriculture (COA). China's organic sector benefited significantly from Beijing hosting the 2008 Olympic Games. The degree of importance of this global event for the organic sector in China can be estimated on the basis of Chinese newspaper headlines and political slogans such as 'Organic food getting popular as Green Olympics nearing' and 'Building the Platform of Organic Agriculture for Serving the Olympic Games' as well as claims in the press of a 50 per cent increase in domestic sales within a year. China's capacity to deliver enough organic food to the athletes and other Olympic visitors was not left to chance: the Chinese government invested in the establishment of highly regulated, strictly supervised and controlled greenhouse production infrastructure for the supply of certified organic fresh fruit and vegetables. The main certified areas are in Northeast China (for grain and bean production), the middle eastern part of Shandong and Jiangsu provinces (for vegetable production) and Southern China – Zhejiang, Jiangxi and Yunnan provinces (mainly for organic tea production). The organic products are mainly exported from these provinces to the EU, USA and Japan.

I Overall policy

In October 2006, the State Environmental Protection Administration (SEPA) issued 'The National Action Plan for Rural Environment Protection' aiming to control pollution and to improve environmental conditions in rural China. Among other things this plan calls for the establishment of an 'organic food production base'. It is estimated that by 2013, 300 such bases specializing in organic production have been set up nationwide as a direct result of this plan. The size of each of these bases varies within a range from only 100 to 10,000 hectares, and each base may cover several villages or towns. The institutional environment for Chinese COA is thus more recent than that for so-called Green Foods (see below) for instance. While organic food for export must meet international standards, domestic organic produce is perceived by policy makers as a complement for non-polluted and Green Food. Certified organic farmers do not currently receive area-based or other 'organic' subsidies, and yet some local governments subsidize the certification costs for producers and processors. Beijing Municipal Government, for instance, decided in 2006 to cover all the certification fees for producers and processors. Xinjiang Province had a similar policy. There are now examples of counties and provinces formulating strategies to increase organic farming and attract companies to establish processing facilities for export-oriented organic products. For instance, since 2004, Zhejiang and Xinjiang provinces have both formulated provincial strategies for the development of the organic sector that cover farming, processing, and marketing. The domestic market remains small, but growing, with significant price premiums in some crops. This is true for both export products (tea and sugarcane) and products such as vegetables sold in supermarkets in Beijing and Shanghai.

There are four levels of food grades in China now: 'organic food', 'Green Food' (minimum of chemicals allowed), 'non-polluted food' (chemicals allowed up to certain levels) and conventional food (minimum food safety requirements). A predecessor to COA, Chinese Ecological Agriculture (CEA) was promoted by the Chinese government during the 1980s and 1990s. Supported with modern science and technology, CEA was seen as an evolution of traditional, biological and organically based agricultural production systems, and a new alternative to decades of conventional agricultural practices (Wu et al., 1989; Cheng et al., 1992). The promotion of CEA, and subsequently COA, involved a development approach based on modern (ecological) economics, stressing their benefits for environmental and rural development (Ye et al., 2002; Shi, 2001; Qu et al., 1997; Jiang and Shu, 1996; Ma, 1988). CEA peaked in the 1990s with reportedly 2,000 pilot schemes and demonstration sites, but it did not fit well with the de-collectivization of agriculture and finally succumbed to supply-side problems and underdeveloped markets (Sanders, 2006, 2000). In contrast, Chinese A-grade Green Food – a product standard – was successful and its logo prevailed across China. Consumers are aware of it and pay a premium for it, generally 20 to 30 per cent above conventional food prices (IFAD, 2005).

In contrast to the Green Foods framework, the institutional arrangements supporting COA evolved only within the last decade. Certified organic production was introduced in some provinces in the 1990s and grew into a large business, mainly driven by export-oriented market chains. An in-depth field study of 15 organic agriculture cases across the Chinese countryside revealed that only six cases arose from market forces – private processing and trading companies contracting directly with farmers – the remaining nine cases involved (county-, provincial- or national-level) government or international agencies (Sanders 2006: 219–24).

II Regulatory set up

The Certification and Accreditation Administration (CNCA) was established in August 2001 with a mandate for national certification and accreditation of different sectors. Since then, China has had a unified regulatory system for organic certification and accreditation activities. Joint implementation is conducted by the relevant ministries and local governments under the overall coordination of the General Administration of Quality Supervision, Inspection and Quarantine of the People's Republic of China (AQSIQ) and the CNCA. Certification rules and specific procedures are jointly formulated by the CNCA, AQSIQ and relevant departments of the State Council, such as the Ministries of Agriculture, Commerce, and Environmental Protection. Examples of the rules and procedures produced by the above system include 'Measures on the Administration of the Certification for Organic Products' issued by AQSIQ, and the 'Implementation Rules of the Certification for Organic Products', and the 'National Standards on Organic Products' (GB/T19630.1–GB/T19630.4 2005), both issued by CNCA. Two different national seals have been introduced covering all organic and 'in-conversion' foods sold domestically. The national standards for organic products have four component parts covering: production, processing, labelling and marketing, and management systems.

The public body initially involved in promoting and regulating COA in China was SEPA, now the Ministry of Environmental Protection, which issued the 'Measures on the Administration of the Certification for Organic Food' (which expired in 2005). SEPA also issued accreditation to certifiers for organic food, when there were few specific policy measures for COA. Authority for standards development, certification management and accreditation was transferred in May 2004 from SEPA to CNCA, an independent body which is not under any ministry, but is attached to AQSIQ. The head of CNCA is usually one of the vice-heads of AQSIQ, which is actually at the same level with other ministries under the State Council.

The CNCA has established an information system about the certification of food and agricultural products. Information regarding certification for organic products is released through the Internet and includes the names of the producer, processor and trader, the issue number of certification, date of expiry, contact person, etc. By the end of 2010, 26 certifiers had received CNCA accreditation and increasing numbers of local inspectors were registered. An estimated 20 per cent of domestic inspectors are trained in China by the International Organic Inspectors' Association (IOIA). The largest certifier of organic products in China, the Organic Food Development Centre(s) (OFDCs), was established in 1994, and undertakes research, inspection and certification of organic foods. OFDCs have more than 20 certified inspectors, of whom 12 have been trained by IOIA. One of the consulting agencies on certified organic products, Dalian Swift Information Consulting Service Ltd., Co., was founded in 2000 and awarded the first organic food consulting agency certificate in China by CNCA. The company has provided certification consulting services to a large number of organic food planting, breeding and processing enterprises in the country.

Some local authorities actively promote certified organic agriculture by subsidizing the certification costs for producers and processors and establishing organic food production bases over several villages. However, with the exception of the period just prior to the Olympic event, there are no all-China national public sector policies of 'buying organic', no overall policies for converting publicly owned lands to organic management and no nationally or provincially recognized Organic Farm Days. A stakeholder consultation undertaken by the Centre for Chinese Agricultural Policy (CCAP) estimates that fewer than 20 per cent of Chinese consumers are aware of the organic label and logo. There is, however, e-commerce for organic produce in major cities and, according to a consumer survey in Tianjing, most interviewees know about Green Food and 'non-polluted' food rather than organic food.

III Research policy

There is evidence of public sector support for the organic sector in the form of advice, training, research and marketing. Organic Food Development Centres support some of the above policy goals. They supply information materials to retailers (supermarkets) and consumers, and since 1997 have sponsored the quarterly journal *Times of Organic Food,* the only Chinese publication about organic agriculture and organic food, which carries news about the development of organic agriculture at home and abroad, experiences of the production, processing and trade of organic food, and introductions to the technologies involved in organic agriculture (www.ofdc.org.cn/products/products.asp). China also has initiatives for organic agricultural research. Agricultural universities and colleges have undertaken agricultural research on organic farming since the late 1990s, when research on organic vegetables was initiated at the China Agricultural University (CAU), which now runs an educational programme on organic agriculture and trade. Organic rice developed by the South China Agricultural University is now exported to Hong Kong. Also Zheijang University does research on organic agriculture. Tea is one of China's major organic products. The Chinese Academy of Agricultural Sciences established the Organic Tea Research and Development Centre (OTRDC) in March 1999, and by 2005

Table 9.1 Organic products of the major producing provinces in China, 2007

Province	Organic product	Certified organic land area (thousand ha)	Domestic sales (million US$)	Export sales (million US$)
Inner Mongolia	sunflower, buckwheat, flax, beans	400	23.5	2
Heilongjiang	soybean, wheat, maize, pumpkin, beans, rice	126	12.9	3
Jilin	soybean, sunflower, gourd, peanut, beans	404	29.9	4
Liaoning	maize, soybean, peanut, wheat, flax, beans	68	66.5	39
Hebei	chestnut, beans, alfalfa, soybean	01	9.7	21
Jiangsu	tea, rice, vegetables	09	62.6	19
Jiangxi	green tea, camellia, rice, strawberry, bamboo shoots	57	27.7	5.3
Fujian	ginger, oolong tea, green tea, fungus	09	5.5	9.5
Yunnan	tea	394	12.5	3.6
Shandong	vegetables, fruit, rice	07	48	21.6

China produced 12,000 tons of organic tea from a total certified area of 16,000 ha (www.tea-trading.com/tea_info/2006_02_20_13_47_15.htm).

IV Civil society

Premiums for organic food are in a range up to 50 per cent. Certification of organic produce for export is done by internationally accredited companies including OCIA (USA), ECOCERT (France), BCS (Germany), IMO (Switzerland), the Soil Association (UK) and JONA (Japan). Local certification is mainly done by OFDC and OTRDC. A number of organic farms near cities, especially those involved in organic fruit and fish farms generate some income from providing tourists from the cities with a choice of activities such as fruit picking, fishing and picnics. Tours to ecological farm households have been established close to several large cities, such as Beijing, Shanghai and Guangzhou. While domestic organic markets do exist, smallholder farmers in many places report difficulties in getting technical assistance and organic inputs, and meeting quality, safety, packaging and labelling standards of traders or supermarkets (*Guangming Daily*, 29 August 2006).

A number of companies in the private sector are active in the production, processing and trade of organic products. One such company, with about 1,200 employees, Yinxiang Weiye in Heze, Shandong Province, produces – among other things – organic dairy and organic feed grass supplied by about 1,300 farm households farming about 1,500 ha. The company provides certified organic milk and yoghurt to retail outlets including supermarkets in Jinan, the capital city of Shandong Province. While a large number of private enterprises engage in organic agricultural production and trade, there is as yet no nationwide sub-chamber on organic agriculture. However, there are a number of national and regional workshops about certified organic production, which take place every year, with the participation of several ministries. The OFDC has organized 13 annual national, and a number of regional, workshops on the techniques of organic production and seminars for exchange of organic information since 1994, when the first national conference on organic farming was held in China. China now has a significant number of active international certifiers and also hosts international donors promoting certified agriculture through various projects and programmes.

In terms of education, Bioasialink – an EU–China co-funded project – has been specifically designed to develop a curriculum on Organic Farming suitable to the Chinese educational framework and to produce teaching materials supporting curriculum implementation, thus strengthening collaboration on education and research between Europe and China 'focusing on Organic Farming as a means to promote sustainable development'. The general objectives of Bioasialink are 'the promotion in China of a better knowledge of the European standards and regulatory framework with respect to the production and marketing of organic food; a network of European and Chinese academic Institutions teaching and doing research on Organic Farming; upgrading of scientific and technical capacity of existing and future teaching staff from Chinese higher education Institutions; the establishment of an educational and teaching platform ensuring long distance education on Organic Farming' (www.bioasia-link.net). A programme on 'Organic farming development supporting system research' was funded by the Chinese Ministry of Science and Technology as long ago as 2004, and from 2005 a research project on organic agriculture strategy development in Western China was supported by the Ministry of Agriculture and marked the beginning of state support for the development of organic agriculture throughout China.

V Context

There is no fully functioning polluter pays principle in operation to provide incentives for additional conversion to organic agriculture beyond its present market or demand-driven niche. There is no tax reduction on inputs for organic agriculture, no support price mechanism for organic products, and organic farming is not yet separately categorized in the otherwise advanced and voluminous body of national statistics. There is no preparation by lawmakers on taxes or tradable quotas for synthetic agricultural inputs and no mechanisms to provide organic farmers with legal redress against contamination by GMO producers or users. Some local governments, however, provide a subsidy to promote 'low poison' and 'low residue' pesticides. In 2011, for example, Shanghai municipal administration provided vegetable farmers with subsidies for buying and applying low poison and low residue pesticides (http://agri. shqp.gov.cn/gb/content/2011-08/25/content_424340.htm).

Comparative analysis and discussion

As summarized in Table 9.2, Brazil and China both have national-level policies and strategies for COA, but to rather different extents. In Brazil federal state rules and three ministries support organic agriculture through a range of policy instruments and development programmes, including some favouring small-holders. Policy rationales include agrarian reform, environmental objectives, food security and rural development. A range of public sector institutions have either initiated policies of buying organic or converting publicly owned land to organic management. In China, COA was initially supported by the Ministry of Environment (SEPA). The state does not provide any specific financial support to organic farmers and as of today it is unclear to what extent public agencies such as the Ministry of Agriculture are committed to supporting COA vis-à-vis conventional farming and Green Foods. However, large areas are planned for conversion to organic agriculture as part of (local) public strategies. Aside from this strategy, support for COA is evident at local levels, where some municipal governments support conversion through reimbursing certification costs and acting as intermediaries between the private sector and smallholder farmers. In Brazil there are also differences in the levels of engagement of regional and municipal authorities and CSOs in supporting organic production. Thus in both countries political leaders and policy documents emphasize the importance of certified organics, and both countries display a degree of regional differentiation.

Elaborate certification or conformity-assessment regimes exist in both countries (see Table 9.2). In China, there is a nationally unified system for organic standards, management and accreditation, but also a well-established product standard for 'Green Food'. In Brazil, the national regulation has not yet been implemented. In both countries, public agencies provide research and education programmes. The development of COA in China has mainly been driven by demand from export markets and the engagement of the private sector (including newly privatized former public agri-food companies) in areas that find it difficult to compete in conventional agricultural products. In Brazil, the opportunity for exporting COA products with a price premium has also been a strong driver for conversion attracting private companies, but this has gone hand in hand with a strong involvement from the public sector and civil society, including locally organized farmers' groups and NGOs which have helped embed organic farming in many areas of the country.

Both Brazil and China have e-commerce of organic produce in major cities, where knowledge about certified organic farming and labels is restricted to

Table 9.2 Summary of comparative analysis

Analytical dimension	Brazil	China
I	Federal state law(s) and the ministries for Agriculture, Environment and Agrarian Development support organic agriculture through a range of policy instruments and development programmes. Policy rationales include agrarian reform, environmental objectives, food security and rural development.	The state does not provide specific financial support to organic farmers and support for COA from public agencies comprises rare cases at low level. However, China's National Action Plan for Rural at Environment Protection envisages establishing 300 organic food production bases, covering between 100 and 10,000 hectares each by 2010. Some municipal governments support conversion through reimbursing certification costs and acting as intermediates between the private sector and smallholder farmers.
II	A diversity of certification and conformity-assessment schemes coexist, including international certification bureaus (CBs) with or without a Brazilian office, national CBs with or gaining international recognition for accessing the main markets and/or accreditation from the IFOAM Organic Guarantee System and CBs that have evolved from organic farmers' and advisers' associations.	Since 2005 a national seal for COA products; a national accreditation and certification body (CNCA) established in 2001. Passed 30 certifiers for COA by 2006. In contrast to COA which involves a process standard, the domestic market is dominated by product standards, including 'Green Food'.
III	EMBRAPA, the national research institute established a programme for organic agriculture in 2000, and in 2002 EMBRAPA launched a project for developing organic agriculture, involving 135 researchers from 15 of its research centres. The National Council of State Agricultural Research Enterprises (CONSEPA) is a consortium of 17 research and development organizations, and has around 40 researchers and advisers directly involved in R&D for COA. There is also involvement in COA at the municipal and state level.	A few university research activities, including the Agro-ecological Research Institute of China Agricultural University, constitute examples that 'organic' research takes place in China. Some of the research conducted on organic agriculture has been funded by the International Fund for Agricultural Development (IFAD), Gesellschaft für Technische Zusammenarbeit (GTZ) of Germany, AMBER Foundation, Greenpeace and the International Institute for Sustainable Development of Canada.

Table 9.2 continued

Analytical dimension	Brazil	China
IV	Strong tendency in Brazil's domestic market towards accepting alternative conformity assessment procedures to complement certification as per international standards. Independent farmers' organizations have played a strong role in the development of organic agriculture and CSOs are involved in organic activities and exert a stronger policy influence in Brazil. Many national, regional and local workshops relating to certified organic production, with multiple stakeholder participation, have been organized. E-commerce of organic produce in major cities.	China has few, if any, classical [civil society] organic farmers' organizations. However national, regional and local workshops relating to certified organic production, involving some stakeholder participation, have been organized. E-commerce of organic produce in major cities.
V	Brazilian policies on organic agriculture play out in a broader context of rural development, food security and health (children's meals).	COA remains poorly integrated with other policy areas such as tax, rural development, and health.

educated segments. Public sector support for the organic sector through advice, training, research and marketing is probably stronger in Brazil, but is also evident in China. In addition the estimated numbers of professional 'organic' agricultural advisers in the national extension service and private sector, and promotion for Organic Farm Days indicate a somewhat stronger institutional environment for organic agriculture in Brazil. Both countries feature a significant number of active international certifiers and both host activities by international donors promoting certified and de facto organic agriculture through projects and programmes. In both countries, small farmers/production units report significant difficulties in meeting quality, safety, packaging and labelling standards.

Organic farming still exists in a 'policy ghetto' vis-à-vis conventional Chinese and Brazilian farming, but to a varying degree. Neither country seems to have operationalized the polluter pays principle – for example in the form of fertilizer or pesticides taxes – as an incentive for promoting organic agriculture beyond its current largely market-led niche. In both countries it makes sense to think the level of embeddedness of organic farming as regionally differentiated, following patterns of regional (and regionally targeted) support and responses to agricultural constraints and marketing opportunities. This seems logical considering the large and differentiated agricultural sector compared with for example European countries. In Brazil, agrarian reform or rural development programmes and civil society might be seen as driving organics deeper into agrarian and rural development policy discourses and measures than in China.

Last, but not least, hosting major global sports is a strong factor in both countries. Our analysis suggests hosting the Olympics was a major factor driving the Chinese government to accelerate investments in the development of the certified organic sector. The 2008 Beijing Olympics demonstrated that the organic agriculture sector could produce a reliable supply of food sufficient in quality and quantity to satisfy the demands of athletes and the requirements of the international Olympic committee. Chinese COA were not developed for the Olympics, but the 2008 Beijing Olympic Games helped spread awareness of green values, and the Chinese expect this trend to continue long after the games are over (www.ebigear.com/restext-1338-7777700042301.html).

Similarly, now, a dozen Brazilian states support organic producers in building capacity to supply certified organic food to meet demand from the 2014 FIFA and 2016 Olympic events (which are expected to significantly increase the market for organic products in Brazil). State agricultural authorities are grasping the opportunity to intensify information on COA and the organic sector law (No. 10,831), partly through the 'Caravan Cup Organic Brazil 2014', which is touring 12 states and 44 major cities, in collaboration with the Ministry of Sports.

Conclusion and policy options

In this chapter, we have presented a framework for analysis of developments and change in five dimensions and at various levels, including state, market, regional and local and civil society. Our analysis confirmed that institutional change is evident at all these levels in China and Brazil, although in different forms and to varying degrees. Indeed, the institutional environments for organic agriculture in China and Brazil presently offer the formal support needed to accredit and certify COA production and thus market COA products in a way that involves a price premium, especially for export. What is not in the making is any Pugliese-style (2001) convergence between COA and sustainable rural development, nor a multidimensional institutional environment sufficiently conducive and embedded to strongly accelerate conversion from conventional to COA, beyond its present niches and towards embracing the two national agricultural sectors as a whole. Therefore while we have demonstrated institutional change, questions remain whether this will eventually bring China and Brazil sustainable rural development involving truly improved environmental outcomes.

Policy option-wise, the comparative analysis in this chapter suggests that:

For Brazil, significant potential exists for:

- Consolidating and expanding the existing domestic policies for developing certified organic agriculture in the domestic and local markets as well as export markets and production landscapes throughout Brazil.
- Developing and expanding the existing EMBRAPA and university research on COA into major and comprehensive research programmes covering all current and future product lines (crops and husbandry).
- Developing the bodies of national agricultural statistics to account for the state of the environment (natural resources and pollution) in conventional and certified organic farming chains and systems.
- Generally improving the conditions for environmentally benign production systems through a strengthening of the incentive structures favouring clean technologies and ensuring the polluter pays principle becomes and remains operational, to provide incentives for additional conversion to organic agriculture.
- Considering quotas for synthetic agricultural inputs and mechanisms to provide organic farmers with legal redress against contamination by resource depleters and polluters, including producers of GMO.

For China, significant potential exists for:

- Formulating a domestic policy for strengthening and integrating certified organic agriculture in the domestic market and production landscape, to complement the existing product standard of 'Green Food'.
- Strengthening, perhaps coordinating and synergizing, existing university research on COA.
- Including COA in the body of national statistics and developing the agricultural statistics to account for the state of the environment (natural resources and pollution) in conventional, Green Food and certified organic farming systems.

- Strengthening incentive structures favouring clean technologies and ensuring the polluter pays principle becomes or remains operational, to provide incentives for additional conversion to organic agriculture.
- Considering quotas for synthetic agricultural inputs and mechanisms to provide organic farmers with legal redress against contamination by GMO producers and/or other polluters.

References

Cheng, X., Han, C. and Taylor, D.C. (1992) 'Sustainable Agricultural Development in China', *World Development,* 20, 8, pp. 1127–1144

Darolt, M.R. (2004) 'Desenvolvimento rural e consumo de produtos orgânicos', in J.B.S. Araujo, and M.F. Fonseca (eds.), *Agroecologia e agricultura orgânica: cenários, atores, limites e desafios. Uma contribuição do Consepa.* Campinas, Consepa, pp. 11–33

Fonseca, M.F., Ribeiro, C. de B. and Vieira, G.Z., II ENA (2006) 'Actividades do grupo de trabalho Relação com os mercados'. Relatório técnico analítico. Rio de Janeiro, GT Mercados ANA, July

Guivant, J., Fonseca de A-C., M.F., Ramos, F.S.V. and Schweizer, M. (2003) 'Os supermercados e o consumo de FLV orgânico certificado', Annex III in M.F. Fonseca de A-C, and C. Relatório final do projeto CNPq sobre harmonização das normas na agricultura orgânica, Projeto CNPq n. 052874/01-3 concluído. Niterói: PESAGRO-RIO

Ho, P. (2006) 'Trajectories for Greening in China: Theory and Practice', *Development and Change,* 37, 1, pp. 3–28, www.ifoam.org/about_ifoam/standards/pgs/pdfs/PGSConceptDocument.pdf (accessed 26 June 2006)

IFAD (2005) 'Organic Agriculture and Poverty Reduction in Asia: China and India Focus', Report No. 1664. Rome, IFAD

Jiang, X. and Shu, J. (1996) 'The Application of Ecological Economics on a Chinese Ecological Farm', *Ecological Economy,* 1, pp. 4–33

Ma, S. (1988) 'More Attention to Ecological Development of Agriculture for a Sound Ecological Balance', in Guo, S., Zhang, W. and Wang, W. (eds), *Ecological Agriculture in China* (in Chinese), Beijing, China Prospect Press

Medaets, J.P. (2003) 'A construção da qualidade na produção agrícola familiar: sistemas de certificação de produtos orgânicos', Tese de Doutorado. Brasília-DF, Universidade de Brasilia, Centro de Desenvolvimento Sustentável

Medaets, J.-P. and Fonseca de A.C., M.F. (2005) 'Produção, C. orgânica: regulamentação nacional e internacional'. Brasília, NEAD, NEAD. Estudos No. 9. Disponível em: www.nead.gov.br/index.php?acao=biblioteca&publicacaoID=314 (Acesso em: 17 ago)

Meirelles, L. (2004) 'Centro Ecológico, Rio Grande do Sul, Brasil', in A.P. Lernoud (ed.) Taller de certificação alternative, 13 a 17 de abril. Torres-RS-Brasil. Guia del taller. pp. 27–33

North, D.C. (2005) *Understanding the Process of Economic Change,* Princeton, NJ, Princeton University Press

Nunes, M., (2004) 'ACS. Pesacre, Brasil', in A.P. Lernoud (ed.) *Taller de certificação alternativa,* 13 a 17 de abril, Torres, RS, Brasil, [Torres: IFOAM-MAELA]. Proorganico.. Programa de desenvolvimento da agricultura organica 2004, pp. 22–27

Pugliese, P. (2001) 'Organic Farming and Sustainable Rural Development: A Multifaceted and Promising Convergence', *Sociologia Ruralis,* 41, 1, pp. 112–130

Qu, F., Kuyvenhoven, A., Heerink, N. and van Rheenen, T. (1997). 'Sustainable Agriculture and Rural Development in China', in van den Bergh, J. and van der Straaten, J. (eds), *Economy and Ecosystems in Change: Analytical and Historical Approaches,* Cheltenham, Edward Elgar, pp. 185–200

Sanders, R. (2000) *Prospects for Sustainable Development in the Chinese Countryside: The Political Economy of Chinese Ecological Agriculture,* Aldershot, Ashgate

Sanders, R. (2006) 'A Market Road to Sustainable Agriculture? Ecological Agriculture, Green Food and Organic Agriculture', *Development and Change,* 37, 1, pp. 201–226

Santos, L.C.R. dos Rede (2004) 'Ecovida de agro ecologia e certificação participativa em rede: uma experiência de organização e certificação alternativa junto à agricultura familiar no Sul do Brasil', in J.B.S. Araujo and M.F. Fonseca (orgs.), *Agroecologia e agricultura orgânica: cenários, atores, limites e desafios. Uma contribuição do CONSEPA,* Campinas, CONSEPA, pp. 159–187

Shi, T. (2001) 'Moving towards Sustainable Development: Ecological Agriculture in China', in Proceedings of the International Conference on Ecological Environment Construction and Sustainable Development, 24–26 May, Wuhan, China: Jianghan University Press, pp. 3–7

Shi, T. (2002) 'Ecological Agriculture in China: Bridging the Gap between Rhetoric and Practice of Sustainability', *Ecological Economics,* 42, pp. 359–368

Songpei, W., Maoxu, L. and Dai, W. (2004) 'The Emergence and Evolution of Ecological Economics in China over the Past Two Decades', Presentation given at the 8th Biennial Scientific Conference, International Society for Ecological Economics, Montreal, 11–14 July

Willer, H. and Kilcher, L. (eds) (2011) *The World of Organic Agriculture – Statistics and Emerging Trends 2011,* IFOAM and FiBL, Bonn, and Frick

Wu, S., Xu, S. and Wu, J. (1989) 'Ecological Agriculture within a Densely Populated Area in China', *Agriculture, Ecosystems and Environment,* 27, pp. 597–607

Ye, X.J., Wang, Z.Q. and Li, Q.S. (2002) 'The Ecological Agriculture Movement in Modern China', *Agriculture, Ecosystems and Environment,* 92, 2–3, pp. 261–281

10 The dynamics and recomposition of agroecology in Latin America

Lucimar Santiago de Abreu and Stéphane Bellon

Introduction

The growth of ecological agriculture in some Latin American countries has been supported or studied by several scholars. Among them, Brandenburg (2002) identified three important phases in its trajectory: (i) emergence of a movement against industrialization of agricultural production; (ii) formation of new groups and social organization forms; and (iii) institutionalization of ecologically based agriculture, accompanied by a partial dilution of its principles. However, a fourth stage of redefinition and recomposition of several alternative agriculture versions can be seen, in which agroecology occupies an important place and influence in agriculture (Ollivier and Bellon, 2011) and rural development (Caporal, 2006). Accordingly, we intend to retrace this process of agroecology recomposition in Latin America, by describing its characteristics and determinants. This is a gradual alternative agriculture regrouping process, under the umbrella of agroecology, whose realm is defended by numerous social players (institutional agencies, social movements and scientific networks).

Agroecology can be viewed as a scientific proposal, a set of practices and a social movement (Wezel et al., 2009). It relies on ecological and social principles and aims to promote changes in the conventional agricultural production process, through an interdisciplinary approach, with the support of participative research work (interaction with rural development players and farmers) and valuing local or lay knowledge. Its future is at stake in this fourth moment. Will agroecology be a new scientific paradigm leading to specific research topics, an alternative way to do research, reconnecting science, nature and society? Will it contribute to sustainable livelihoods, especially for family farmers, or be a cornerstone for the development of agricultural enterprises? Will it be appropriated by social movements to support agricultural and rural transitions? To examine these pathways, the Latin American experience is particularly relevant since it somehow originated the development of agroecology, including combinations of the various dimensions of agroecology.

In order to contribute to answer these issues, this chapter includes five sections. After describing our approach, we show that family farming is socially and economically important in Latin America, with part of the certified production originating therefrom. We stress the need for taking into account local conditions and differentiated groups of family producers characterized by both social and cultural differences and diversity of ecologically inspired production styles. Another aspect dealt with is the history of alternative agriculture from the 1970s, with identification of its successive events and characteristics, and the dynamics of alternative agriculture into the building up of agroecology and its different features in Latin America. Then, we analyse the current status of agroecology structuring in Latin America, i.e. the elements favoring the setting up of research and development programmes, as well as the participation of social movements.

Research approach and assumptions followed

This work aims at qualifying the meaning and relevance of agroecology in Latin America, when applied to agricultural and food systems. It is based on an interpretation of the theoretical field of agroecology, its practices and attached knowledge, its achievements, its institutional and political commitments. Our approach relies basically on a review of literature and documents, combined with a multidisciplinary (sociological and agricultural) analysis of material developed from situations studied by the authors.

The setting up of training programmes for extension agents and researchers (Master's and PhD programmes and enablement of producers) in several Latin American regions is worth mentioning. Also, the role of the University of Cordoba (Institute of Sociology and Rural Studies, ISEC) in the formation of research and new Brazilian research and development programme directors should be stressed, as shown in the extensive Spanish and Portuguese literature and as observed in recent congresses held by the Brazilian and the Latin American agroecology associations (Sarandon and Flores, 2010).

A scientometric analysis based on 335 references from the Web of Science (Ollivier and Bellon, 2011) showed that most of the academic production in agroecology comes from the USA, also in collaboration with Latin Americans (Figure 10.1). Brazil and Mexico appear among the ten major publishing countries, based on the institution of the first author. Albeit based in Berkeley (California), Miguel Altieri is very influential in Latin America and the major contributor to articles or books (n = 28) where the term agroecology or agroecological appeared in the title or keywords (Wezel and Soldat, 2009).

Due to the English bias of the Web of Science, the previous analysis does not take into account many publications in Spanish and Portuguese, as also noticed by other authors (Tomich et al., 2011). For instance, a first analysis conducted in 2008 (Venturelli et al., 2011)[1] revealed the number and sources of texts written in Latin America, e.g. (i) *Revista Agriculturas*, *Revista Brasileira de Agroecologia* in Brazil and *Revista Latino-Americana de Agroecologia*, among journals dedicated to studies in this subject area, (ii) papers presented at Latino-American and Brazilian Congresses of Agroecology, as well as (iii) numerous postgraduate theses.

Our initial assumption is that these different ecological agriculture styles and empirical practices in Latin America may be related to theoretical conceptions

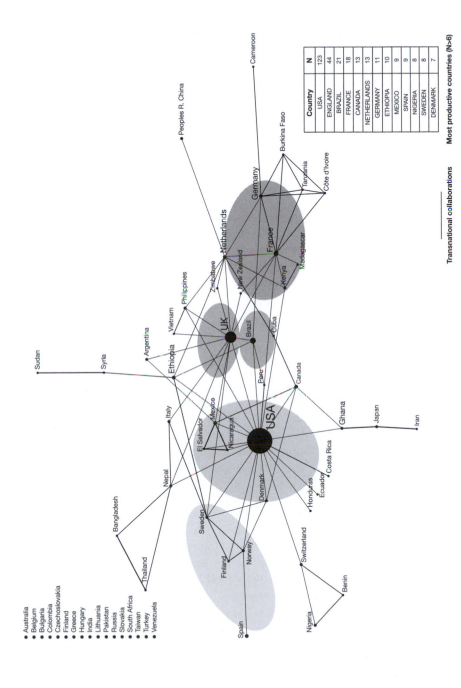

Country	N
USA	123
ENGLAND	44
BRAZIL	21
FRANCE	18
CANADA	13
NETHERLANDS	13
GERMANY	11
ETHIOPIA	10
MEXICO	9
SPAIN	9
NIGERIA	8
SWEDEN	8
DENMARK	7

Most productive countries (N>6)

Transnational collaborations

Figure 10.1 Country affiliation of academic production and collaborations in agroecology

Source: Tichit et al., 2009.

that influence local agents and players, in a context in which scientific and social networks dedicated to agroecology are evolving.

The development of agroecology has to do with producers, government technicians, researchers, consumers and the society's interested groups, as well as the municipal, state and federal governments, among others. Against this background, agroecology as a social project for modern, environmentally friendly agriculture tends to epitomize agrarian issues related to the various social forms of family production. This is because it creates a favourable environment for solving crucial problems connected with employment and (internal) migrations, the dominant food implications and the preservation of biomes. Also, the development of agroecology may be contributing to rethinking the conventional agriculture model through the disclosure of successful initiatives. Accordingly, our position is that the development of agroecology depends on the interaction of social movements, scientific networks and agroecological practices (Agrisud International, 2010) with the ability to create public policies.

Empirical context and transition processes for ecologically based family agricultural production in Latin America

In discussing family production and its specificities, this section stresses the economic and social importance of family production for the development of agroecology.[2] Note that in the production regime presented, diversified social forms of ecologically based production and different agricultural styles are found, with various knowledge frameworks, namely biodynamic, natural, organic, permaculture, agroforestry,[3] etc., which are convergent with agroecological principles, to some extent. See for example Chapter 4, where specific aspects of agroecological methods are put in relation to organic agriculture.

Latin American official statistics are still scarce, and incomplete (with ill-measured organic production – even non-existent in some countries), and with low frequency of data-gathering on production structure and socio-economic indicators. For instance, the latest Brazilian agriculture survey was conducted in 2006. And so far, limited specific statistical data were available on organic agriculture. The latest census was an attempt to identify the organic production systems. Organic producers' farms totalling 90,497 ha accounted for approximately 1.8 percent of all farms studied (IBGE [Brazilian Institute of Geography and Statistics], Censo Agropecuário, 2006). However, the research methodology included in the total number of organic producers those who simply do not use agrotoxics (Abreu et al., 2009b; Willer and Yussefi, 2008, based on a survey conducted in 23 Latin American countries). This does not reflect transition perspectives in agroecology.

From the perspective of a transition towards a more sustainable agriculture, various authors (Hill and MacRae, 1995; Rosset and Altieri, 1997) differentiate three approaches: (i) increased efficiency of input use, (ii) input substitution or the replacement of agrochemical inputs by environmentally more benign inputs (e.g. botanical or microbial insecticides), and (iii) system 'redesign' arising from the transformation of agroecosystem functions and structure. Gliessman (2007), also active in Latin America, adds a fourth level of transition, consisting in re-establishing a more direct connection between those who grow the food and those who consume it. This leads us to consider another system, the agri-food system, beyond agroecosystems. According to these authors, the prevalence of input substitution drastically limits the potential solutions to the socio-economic and ecological crisis of modern agriculture, in the sense that this substitution does not call into question monoculture or the dependency on external inputs. In the Brazilian state of Rio Grande do Sul, these paradigms helped to classify all farmers into three categories: conventional farmers (who reduce their use of chemical input for economic reasons), farmers in transition (who substitute chemical inputs with biological ones) and farmers in a 'redesign' type of transition (with an ecological and systemic approach incorporating soil ecological management, crop rotation and diversification, mixed crop–livestock integration, reforestation and management of agroforestry systems). To assess this transition process, our analysis assumes a dual approach. The first one is descriptive, comparing farmers' actual practices with agroecological principles (Guthman, 2000), or with organic standards as proposed by Darnhofer et al. (2010) about the conventionalization thesis and applied in various contexts by Oelofse et al. (2011). Where applicable, these practices can also be compared with conventional farmers' practices, considered as a reference. In the second approach, we use proxy indicators for each of the three previous categories, where agroecological indicators used are adapted according to the level of transition considered. In any case, the reference framework used to address transition issues is at stake. The necessity of a broad approach for such investigation is also illustrated by the livelihood approach presented in Chapter 1, and complex challenges faced by agricultural production systems, e.g. climate change (Chapter 5) or power relations (Chapter 6).

From a sociological point of view, we highlight the importance of a conceptual meaning of the denomination 'ecologically based social forms of family production', which presupposes the existence of different transition and relation-with-natural-resources models including separate socio-cultural profiles and economic conditions, in turn pointing to dynamic courses and different forms of evolution. This concept was applied by Lamarche (1999) and a team of researchers to study family production in several countries. They noticed that in every country in which changes are organized by the market, to a certain extent the farm production is always guaranteed by producers, among which families stand out. Partially present all over the world, whatever the country, its history and political regime, socialist or capitalist, and whether industrialized or under development, this social form of production in particular evidences a great adaptation capacity. That is why – far from disappearing, as has been foreseen – family agriculture is currently taking a comprehensive, and more universal, dimension. Such a recognition also relates to the specific relationships between family agriculture and environment (Abreu, 2006). We redefine the concept and adjust it in the light of the ecologically inspired agriculture conditions.

Box 10.1 Data on certified organic and agroecological production in Latin America

The land cultivated by certified organic agriculture in Latin America is used for a highly diversified range of food products, such as honey, vegetables, coffee, sugar, palm-tree fruit, soybean, eggs, meat, milk and corn, among others. Some of the markets are sufficiently developed to show that we are no longer in 'niche markets'. The biggest Latin American producers, Brazil, Argentina and Uruguay, alone cultivate over 4 million ha. Argentina owns 2.8 million ha, of which 2.3 million ha are used for cattle-raising activities. Costa Rica also stands out in the organic production field, especially in connection with coffee. Currently (2012), 6.4 million ha of the total area are estimated to be used by certified organic production systems. Another 6 million ha are used for sustainable handling of agricultural produce under some kind of certification, e.g. forestry certification, as in the case of Brazil, Costa Rica, Argentina and Uruguay. Brazil has 887,637 ha of certified land and 5.9 million ha under forestry certification.

Research conducted in Latin America shows that there are differences in the profiles of the countries under study, regarding areas used by organic systems and the number of certified production units (Willer and Yussefi, 2007). Despite owning the largest area, Argentina has a small number of production units, with only 0.27 percent of certified area in relation to the total arable land in the country. In countries like Peru, Cuba, Costa Rica, Honduras, Guatemala and Mexico, production by family producers working in smallholdings predominates, whereas Argentina, Colombia, Venezuela, Panama and Chile have small- and medium-sized properties, given the large extent of land and the small number of production units. In Brazil, 20 percent of the organic production systems is formed by very different social forms of production, namely family-run businesses: family production units operating in various types of market; or family subsistence activities, using smallholdings or no land at all, and renting other owners' land. In this category are over 15,000 farmers (0.3 percent of the total). Of the Brazilian certified production, 70 percent is directed towards the international market (sugar, orange juice, coffee, soybean, cocoa, yerba mate, palm oil and dried fruit, cashew nuts, guarana, etc.) (Abreu et al., 2009b). A study in the Andean community member countries showed that in Bolivia, 67.2 percent of family agriculture is subsistence farming, in Colombia the percentage rises to 79.4 percent, in Ecuador it is 61.6 percent and in Peru it falls to 45.5 percent. These ecologically based social forms of production are characterized by family insecurity, little land available, no access to credit and insufficient income. The remaining agricultural activities are divided into socio-economic transition family agriculture and business-like family agriculture (Tello, 2011). In the Andean countries extreme efforts have been made to institutionalize family agroecological initiatives; such as in Ecuador, within the 'Conference on Food Sovereignty' sphere of action, and in Colombia with commendable political willingness on the part of the Ministry of Agriculture and Rural Development. In the case of Bolivia, the movement is guided by the motto 'A Bolivia Digna y Soberana' (Dignified, Sovereign Bolivia) resulting from a partnership between the private sector and civil society organizations. In Peru, the Ministry of Environment has enthusiastically supported this type of initiative by planning sustainable natural resource management actions as well as those taken by organic production system farmers (Tello, 2011).

Organic agriculture in Latin America has grown both in the domestic market and for exportation. The main export products are coffee and bananas from Central American countries, sugar from Brazil and Paraguay, and soybeans from Brazil and Argentina. Coffee is one of the main export products. Income from sales of organic coffee is earned by approximately 500,000 people. The 'Bolivian Association of Organizations of Ecological Producers' (AOPEB), founded in 1991, already has 30,000 members. Its programme is intended to strengthen and socially integrate 32 organic coffee producers' associations to meet the local and foreign market demand.

Another important product in Latin America is cocoa, with total production of 2.7 million tons a year, and consumption rising by 16 percent to 20 percent each year on average. The Dominican Republic is the biggest producer and exporter of cocoa, with a production of some 14,350 tons a year, 60 percent of which – 8,500 tons – is sold. The total organic certified cocoa production is 57,000 tons, 98 percent of which is produced by Latin American countries. The total organic cocoa production area in Latin America is 73,000 ha. The main producers are the Dominican Republic, Mexico, Ecuador, Peru, El Salvador and Brazil.

Brazil, Paraguay, Ecuador and Argentina are the biggest organic sugar producers. In Argentina, in the Misiones region, the San Javier Organic Cooperative includes 650 members who process 80,000 tons of sugar a year. In Paraguay, around 1,000 organic sugar producers earn 20 percent more for their produce, than the price of conventional sugar.

The domestic markets of Latin American countries differ from one another. In some of them sales to supermarkets and hypermarkets predominate, whereas in others, sales are made to small- and medium-sized markets. However, in both cases direct sales are a constant (at street markets and by home delivery). As from the 1990s, the supermarkets have opened up to sales of organic products, especially vegetables and fruit. In Brazil,

Uruguay, Argentina, Costa Rica, Peru and Honduras, initiatives of this kind are numerous.

Organic producers in most of the Latin American countries are family producers working with the previously mentioned small- and medium-sized agricultural structures. In general, they are associates or members of, respectively, a cooperative or a social organization, and local markets have been the most feasible alternative to guarantee sales of their produce.

In Brazil, an example of this type of sales is 'Rede Eco Vida', which represents a formal recognition of agroecology through a Participatory Guarantee System (Bertoncello and Bellon, 2008). This network, which covers the Southern Region states, promotes weekly street markets in several cities, including the capital cities. A similar example is found in Ecuador, where 'Fundación Maquita Cushunchic' also promotes direct producer–consumer sales. In Costa Rica there is a slogan that helps increase direct sales: 'From my family to your family'. In Brazil, ecological street markets have existed since 1989 in several towns in the states of Rio Grande do Sul, São Paulo and Paraná. The street markets are organized by groups of farmers themselves. The number of consumers involved varies between 100 and 10,000. A big international trade fair of organic products – BioFach – is held in the city of São Paulo. Other examples are in the 'Agroecological Network' in Peru, with its weekly street markets in several towns in the interior, faithful to a millenary Peruvian Indian community's custom. In Uruguay, at Parque Rodó, there is also a weekly street market and in the Dominican Republic an ecological market – FAMA – exists in the capital, Santo Domingo. Similar initiatives can be found virtually everywhere in Latin America (Tassi, 2011).

Some case studies on production strategies and the application of agroecological principles revealed that the establishment of economic relations between organized groups (associations, cooperatives, etc.) and trade entities (fair trade was one of the cases studied in Brazil) has stimulated a significant increase in production diversity and application of ecologically inspired agriculture principles as well as noticeable change in the attitude towards exploration of natural resources at several locations (Almeida and Abreu, 2009).

In Brazil the institutional point of view on organic agriculture (Brasil, 2003) was inspired by agroecology. Both systems advocate similar principles and common practices, except for the market and certification. The institutional vision of organic agriculture aims to meet the demands of the international market, and not only the local one. However many of the militants who advocate agroecology also consider that organic agriculture would be centred in input substitution and aims to occupy spaces in supermarkets and international markets. Such a critical view denies the contribution of organic systems to redesign and transition. Organic agriculture can also be seen as a broad notion that connects various forms and models of ecologically based production, expresses different situations of transition processes, and will also result in the adoption of a number of the principles advocated by agroecology (Abreu et al., 2009a). There is no general consensus on this position, in particular if we consider the history of the movement of alternative agriculture and the whole geographic area of Latin America. The agroecological approach was more strongly disseminated in Latin America, and largely influenced the process of institutionalization of organic production. This may be one of the reasons for the differences in design and vision between the global North and South, but also fluidity exists between organic agriculture and agroecology. In developed countries the differences are more striking (Bellon et al., 2011). Therefore, we must take into account the differences between continents to understand these different interpretations of agroecology and other models based on ecological production.

So, the family agriculture contribution to the society's food security is undeniable (De Schutter, 2010). The importance of this social segment to food security exceeds primary production. The way income is distributed and employment is created enables millions of people to access food, and this has similarities with other Latin American countries, especially Argentina and Uruguay.

Agroecology – history and evolution in Latin America

In retracing the trajectory of agroecology in Latin America, in this section we stress the emergence of this process and the gradual regrouping of alternative agricultures under the umbrella of agroecology (Abreu et al., 2009).

The term agroecology was first used in two scientific papers early in the twentieth century (Gliessman, 2007; Wezel et al., 2009). However, it was only from the 1970s, concurrently with criticisms of the Green Revolution, that agroecology began to be seen as a scientific discipline. According to Gliessman (2007) and Wezel et al. (2009), early in the 1980s agroecology began to evolve gradually as a social movement and a set of practices. The agroecological movement comprises not only a group of farmers working towards food security, sovereignty and autonomy, but also social movements campaigning for public policies committed to the application of agroecological principles. Also, it may be a movement of farmers formed to respond better to ecological and environmental challenges in the face of highly specialized agricultural production, as in the case of the United States (Wezel et al., 2009).

Associated non-governmental organizations (NGOs), especially those linked to the Catholic Church ('Comunidades Eclesiais de Base'), trade unions, environmentalists and agriculture professionals, were responsible for the first initiatives and emergence of ecological agriculture. They aimed at supporting small farmers who were in a bad situation and marginalized by the process brought about by the Green Revolution, new producers from the big cities, often young environmentalists (movement of appreciation of the countryside; Karam, 2001; Brandenburg, 2002). For some authors, the origin and development of ecological production systems lie on traditional Andean systems, which in turn derive from local knowledge of how agroecosystems function, as for example in the Andean countries (Altieri and Toledo, 2011), and the Brazilian Amazon (Abreu and Watanabe, 2008). This raises the issue of the application of agroecological methods in other environments, especially for landless farmers ('assentamentos'), whose farm trajectories vary among livelihood both before and after their settlement. Interestingly, most of the social movements in Latin America also refer to the scientific dimension of agroecology.

Particularly in developed regions, ecological agriculture was initially known as alternative agriculture (Warner, 2007). The alternative agriculture concept in Latin America reflected a set of techniques to be used in an integrated manner and in balance with the environment. Interestingly, the term ecological is used for organic agriculture in Latin American countries. And many of the organic founders (Besson, 2011) also inspired both organic (Sixel, 2003) and agro-ecological (Altieri, 1987; Altieri and Nicholls, 2003) developments.

The alternative agriculture concept was gradually influenced by agroecology (Ollivier and Bellon, 2011). The idea of an agricultural system focused on alternative techniques lost ground due to the cultural disruption in the core of the environmentalist movement. This movement led by the NGOs and professionals from the field split up due to the criticisms of the production model conveyed by the Green Revolution or industrial agriculture, known as a package of agricultural and cattle-raising technologies characterized by little in the way of practical, ecologically inspired experiments (Almeida et al., 2001).

In this context of agroecological programme identification and building-up, the NGOs played a decisive role. Organizational structures were created in several Latin American countries, with a substantial increase in the number of projects. In the 1990s, the scenario was one of evolution and redefinition of methods and concepts. As a result the term agroecology was introduced in Latin America in the late 1980s against a background of articulation and cooperation between alternative projects. In 1989, the 'Latin American Agroecology and Sustainable Development Consortium' (CLADES), currently the 'Latin American Sustainable Development Center' was created[4] (Altieri and Toledo, 2011). Early in the 1990s, during a big meeting organized by CLADES, with the participation of a dozen NGOs, the idea of agroecology was presented by Miguel Altieri. According to Peterson (2007), the idea of technology transfer was substituted with one of social agroecological innovation. Technology is not an external thing, but notably the result of ecological and socio-cultural relations.

In Latin America, food production based on agroecological principles has grown over the last 20 years, and in some regions, as in the case of Brazil, the origin of this type of production can be partly interpreted as inherited from European models (Brandenburg, 2002) and adapted to the Brazilian context, while coexisting with other original forms established under specific local conditions (Bellon and Abreu, 2006; specifically for the organic sector in Brazil, see Chapter 9). The diversity of ecological production types results in a dynamic mosaic, under full development in the continent. In other words, agroecology is increasingly turning into an intermediate element capable of reuniting the various social, ecologically inspired production forms.

The reason for adopting the agroecology propositions is now a series of actions taken by political organizations committed to the building of a new society model, based on claims for equality and social justice. It is the family producers' standards of living coupled with ecological and technical factors that configure the dynamics of the emergence of agroecology, thus leading farmers concerned about certain situations of land-use transition into differentiated trajectories. Farmers from Brazil, Peru and Bolivia are supported by organizations directly involved in rural development processes and institutionalization is in course in several countries. They fight for recognition of these ecological production forms, representing a social force that is one of the fundamental components of an intensive evolution observed before 2003. These advances end up embodied into an institutional picture, specifically in the case of Brazil, where institutionalization has been heavily influenced by the agroecology concept. The pertinent legislation recognizes the importance of cultural integrity of rural communities, social equity, valuation of family production, and respect for natural resources (Almeida, 1988; Bellon and Abreu, 2006; Bertoncello and Bellon, 2008).

Various facets of agroecology

The theoretical contributions to agroecology are heavily influenced by agronomy, ecology and sociology. Agroeocology is regarded as an emerging scientific proposition and a transdisciplinary field of knowledge (Dalgaard et al., 2003; Buttel, 2003; Ruis-Rosado, 2006). It studies agroecosystems or agricultural units and their transitions on a comprehensive basis, in which mineral cycles and energy transformation occur, taking into account social, economic and cultural relationships. Within agroecosystems ecological processes take place, such as nutrient recycling, animal and insect (fauna) interaction, competition, commensalism and ecological succession (flora). However, the degree of resilience and stability of agroecosystems is not determined solely by environmental and biotic factors, but also by social and economic ones, such as public policies, innovative technologies, land ownership, and markets, i.e. other elements that interfere with the possibility of producers putting agroecology principles into practice. There are two broad categories: (i) descriptive and comparative agroecology, based on various disciplines (agricultural sciences, social sciences, natural sciences; agroecosystem analysis and landscape ecology; cultural anthropology; ecological economics; political ecology), (ii) applied agroecology (design of agroecosystems, agroecological technologies, ecological pest management, ecological soil and water conservation).

The fundamental principles of agroecology are as follows:

- recycling nutrients and energy on the farm;
- integrating crops and livestock;

- knowledge-intensive methods, based on techniques developed on the basis of farmers' knowledge and experimentation;
- application of the holistic concept of redesign, taking a long term perspective for the evolution of farming systems and transitions in agriculture;
- production-oriented for food sovereignty and social justice, including a crucial question associated with the right ethics of proper and sufficient nutrition;
- sales-oriented to meet the demand of local markets and the establishment of direct relations between producers and consumers.

So the concept of agroecology is broad to include systems design issues associated with ethical values, and human and political demands for change. Many of the above principles also differentiate agroecology from eco-agriculture, evergreen agriculture, conservation agriculture, ecological intensification or enhancement, among other eco-logically based modes of production. These principles are the result of various contributions, which we now explain, with a historical, socio-economical dimension and regional case studies.

The origin of the agroecology conceptual matrix is closely linked to three universities – two in California, USA, and one in Andalucia, Spain. The University of California, Berkeley is directly connected with Miguel Altieri's contribution and the University of Santa Cruz to Stephen Gliessman's original work (Sauget, 1993). The contributions made by the University of Cordoba (ISEC)[5] should be stressed. According to ISEC's director E. Sevilla Guzmán (personal communication), the change in the paradigm was due to the introduction of a few natural science elements to explain the nature of industrial agriculture and its impact as compared to natural resources and society itself. This school of thought introduced the agroecology concept and the socio-political sense of agricultural development including its historical processes. This approach leads to a critical view of the neoliberal and global development model. Accordingly, in searching for solutions for the various forms of socio-environmental decay, inquiries into the duality of science (the relation between epistemology and power structure) were stimulated, with concurrent investigation and intervention, always stressing the need for recognizing and prizing local peasants' and indigenous groups' knowledge. Subsequently the local knowledge associated with management and maintenance of natural resources was integrated within the agroecology concept, with the attendant appreciation of this knowledge as science. For many of these precursors, the empirical knowledge is called traditional, for others, lay or local knowledge.

As a scientific field, agroecology shifted from the study of agroecosystems to ecological food systems (Francis et al., 2003; Gliessman, 2007). From this standpoint, producers and consumers are directly interconnected. Consequently, this definition of agroecology may expand the field of social and political evolution. With the integration of ecology in the whole food production system, it also includes social and institutional relations connected with production, and the distribution and consumption thereof. Food systems are also expected to enhance biodiversity, resilience, energy efficiency and social justice (Kerber and Abreu, 2010). Producers' culture and traditions relating to land use and management are considered as strategic assets, in terms of energy, production and food sovereignty (Altieri and Toledo, 2011). According to Wezel et al. (2009) three dimensions are present in agroecology, there being interactions between the following: political vision (social movement), application of innovative technologies (practices) and creation of knowledge (science). Just as in European countries, these interactive dimensions cannot be observed in all Latin American countries with the same intensity.

The key agroecology concept is the agroecological transition expressed by the idea of redesign (Gliessman, 2007). The great majority of Latin American authors, though involved in rural development, refuse to accept that agroecology is often presented as an alternative production method (Caporal and Costabeber, 2004). Agroecology also has a critical position in relation to organic agriculture (Altieri and Nicholls, 2003), although since 2003, both of them are grouped together in a single institutional mechanism that regulates production and the market (Brasil, 2003; Bertoncello and Bellon, 2008). The criticism remains and above all is centred on the minimalist view of organic agriculture, in as much as it is seen as a simple input substitution within a production context dominated by family production, to the detriment of a redesign of agricultural systems. Despite the criticisms, the conversion to organic agriculture is often used to illustrate the prospects of agroecological transitions. As a result, the relationships between organic agriculture and agroecology seem to be rather complex (Bellon et al., 2011), and agroecology may indeed occupy an important place in terms of physical survival and biodiversity valuation.

Agroecology development strategies in Latin America

In order to understand the dynamics and growth seen since the beginning of the 90s, we highlight the agroecology development strategies in Latin America. Several university postgraduate courses in agroecology have been created, and hundreds of NGOs have used agroecology to foster sustainable agriculture. Such programmes benefited from cooperation between Latin American and European researchers with experience from previous projects, developed under a Master's programme called 'Agroecology and Sustainable Development in Latin America and Spain', an initiative of the International University of Andalucia. Its postgraduate (doctorate) programme from Cordoba University (ISEC) was strongly reinforced by interrelations between groups of researchers from various countries, especially in Latin America. According to Sevilla Guzmán (2002) three main objectives favoured this interaction: (1) training and investigation; (2) actions with social movements more directly linked to the landless people movements in Brazil (MST); and (3) research into alternative agriculture models, developed with the collaboration of researchers and rural extension technicians also in Brazil.

This group of researchers and extensionists still maintain strong collaborative relations, and ISEC asserts itself as a scientific research institution intensively staffed by young students and professionals who, in turn, maintain relationships with social and agroecological movements. With the election of Lula as President of the Republic of Brazil, this group of researchers, jointly with the Workers' Party (PT), began to occupy important political positions in the Ministry of Agrarian Development (MDA), among others, and this enabled them to directly influence public programmes, especially in the rural extension and agricultural and cattle-raising fields.

These rural development policies are established through articulation with NGOs and other independent groups. Some institutions initially participated in the so-called alternative agriculture, for example, in Brazil, the 'National Articulation Agroecology' (ANA), created in 2004. So, it also became possible to bring all the interested parties into agreement, with the formulation of the main structuring bases of the 'National Agroecology Research Program' of Embrapa (Empresa Brasileira de Pesquisa Agropecuária) and the 'National Innovation and Systems Research Program' integrating research efforts into organic production, permaculture or agroforestry systems. Public institutions from Brazil, Cuba, Venezuela, Bolivia and Peru have incorporated the agroecology perspective into their rural development strategies. Recently, rural movements (Via Campesina, MST, Movement of Small Producers (MPA), etc.) adopted an agroecology proposition, trying to include in their agendas the food sovereignty theme (Luzzi, 2007). Currently (2012), there does not seem to be a single organized space in the agrarian sciences (professors, researchers and extensionists) in Latin America without professionals committed to agroecology. They are professionals seeking knowledge of alternative and innovative technologies from an environmental point of view, which are appropriate for family producers. Their interest is also in local alternative markets. This knowledge, from an agroecological point of view, depends on a strong interaction between technical, scientific knowledge and local experience gained by organizations of ecologically inspired producers.

The Latin American movement in defence of agroecology is designed to influence the establishment of policies capable of stimulating food sovereignty and sustainable rural development. Given the lack of formal knowledge provided by scientific investigation institutions, the formal education and rural extension and development institutions strive to fill these gaps, by making available to professionals and farmers a set of educational and investigative processes. One of the institutions that has endeavoured to solve this issue is the Latin American Scientific Society of Agroecology (SOCLA).[6] It works jointly with non-governmental institutions and they are supported by social movements, with a view to promoting the development of a scientific basis for agroecology. The creation of scientific knowledge has been defined as strategic and a privileged mission. The major themes addressed are as follows: food sovereignty, conservation of natural resources and biological diversity in agriculture (Letter of Congress of Agroecology SOCLA, 2011).

In order to accomplish its mission, SOCLA organizes a scientific congress every three years, basic courses in several countries and publications on fundamental subjects. Also, it maintains technical task forces, aimed at assisting interested parties and organizations of farmers involved in agroecology in the continent. One of its supplementary initiatives was the creation of a postgraduate programme (doctorate) in Medellín, Colombia, in 2007. At the congress held in Medellín Colombia in August 2007, several task forces were created to participate in research, education and divulgation of contemporary problems affecting Latin America (climatic changes, emergence of biotechnology and biodiesel cultures, impact of globalization and free trade agreements, food sovereignty, etc.), as well as make an in-depth analysis of the status of agroecology in several scientific fields such as soil management, pest management, sustainability indicators, ecological economics, ethnoecology and rural development

Accordingly, agroecology tends to become a reference for changes in production methods, by trying to reconcile development and the interests of society, such as food security and empowerment of rural populations. The results of these research efforts are channelled to the organized social movements, thus improving the activities and facilitating the development of technological alternatives, fairer distribution systems, rural development strategies on a local basis, and changing policies that favour sustainable food systems.

The great challenge seems to be coexisting with a production-intensive agriculture, as currently there is an intense movement towards occupation of large areas of arable land for growing biofuel-generating plants. However, no scientific study has been conducted to determine the impact of these events on food security and biodiversity, in the region. Likewise, in Embrapa researchers have developed models to foresee the impact of climatic changes on agricultural productivity, although studies on how to render agroecosystems resistant to drought, climatic events or irregular rainfall patterns are scarce.

The following are the most important collaborative links among several non-governmental organizations in Latin America: the 'Latin American Agroecological Movement' (MAELA); organized social movements (Via Campesina, MPA/Brazil, ANAP/Cuba, ANPE/Peru, etc.). MAELA is composed of NGOs, family producers' organizations, consumers, indigenous populations, agroecology movements and networks, and educational institutions and universities. In its 16 years of existence, MAELA has attracted approximately 250 organizations in 20 countries. It is a reference in the ecological agriculture field. Once more, the relevant themes and those dealt with by SOCLA seem to converge: food sovereignty, transgenic products, biodiversity, agricultural research and legislation regulating organic production systems. One of MAELA's objectives is the socialization of experience in 'Participatory Guarantee Systems'. Also, it stimulates free regional circulation of ecological products, with due regard for legal restrictions. Currently, MAELA is divided into three regional parts: Mid-America, Andean Region and Southern Cone. These three regional groups work to spur family producers' sales at local street markets and trade fairs, to increase their income and strengthen organizations and cultural centres. They also focus on preservation and appreciation of local knowledge and access to healthy, fresh products, all in the hope that the results may provide farmers' organizations with the means to establish production policies and create innovative markets (Relatório de viagem/Reunião no Peru; see also Tassi, 2010).[7]

The IFOAM's Latin America and Caribbean Group (GALCI)[8] has strived to strengthen the Latin American organic movement. It supports transfer of knowledge of, and experience in, domestic and international markets. Their objective is also to stimulate the organizational capacity of family producers in the continent. The interconnection of networks has resulted in a series of meetings, with the production of important bibliographical material as a result. The subject has already been addressed at a meeting in Porto Alegre, on the theme 'The state-of-the-art in Agroecology'. Latin American and Spanish researchers, technicians, and students from Cordoba University participated in a symposium, thus enabling a profitable exchange of information and close work relations to be established.

In sum, the agroecology development in Latin America has been guided by political strategies adopted by the various players involved and an increasing interest on the part of society as a whole and consumers in particular. Scientific interdisciplinary programmes, agricultural practices and social movements have been included in the development process. The agricultural practices and scientific programmes are built on the basis of knowledge derived from interaction of technicians, researchers and the various ecologically inspired players located in the different Latin American regions, who are involved in local agroecological experiments. Therefore, it intends to combine a diverse knowledge and disciplines.

Altieri and Toledo (2011) suggest that Latin America is conducting an unprecedented agroecological revolution. They refer to the in-depth changes in the social, epistemological and technological fields. This use of the term revolution seems highly questionable, in that it would be the consequence of political willingness and institutional, professional and scientific structuring. In fact, agroecology develops, evolves and coexists with an exportation model. At the same time, the authors indicate that this revolution depends on the peasants' access to land, seeds, water, credit and local markets, partly through the establishment of economic support policies, financial incentives, market opportunities and agroecological technologies. The development of agroecology in Latin America may be restricted by the growth of production-intensive agriculture and exports, with their own structures and financial support from the government. In the case of Brazil, under the new 'Crop Plan' 2011/2012 R$107 billion was budgeted for agribusiness and a further R$16 billion for family producers. It is worth mentioning that the conventional, government-supported agricultural sector has also increased production, in the interest of economic groups represented by employers (grain and fuel producers) and exporters. This production model has been fiercely criticized because of its negative impact on the environment and social issues, its omission of human health issues, taking into account the harmful effects of agrotoxics used in food production, and the fact that Brazil is one of the biggest consumers of agrotoxics in the world. Therefore, we are far away from a revolution per se.

Target groups and up-scaling agroecology remain controversial. Some authors (e.g. Altieri, 2002) consider small farmers excluded from agricultural modernization as target groups for agroecological transitions. Other authors (Warner, 2007) suggest that agroecological practices can also concern the heart of industrial agriculture, improving product quality while avoiding environmental impacts. In Brazil, two visions are confronted (Beauval, 2010): one is advocating conservation or no-tillage on large and mechanized farms, to reduce production costs and improve soil fertility while using herbicides and simplifying crop rotations; the other one is geared at strengthening the autonomy of small farmers who are valuing agrobiodiversity, reducing the use of external inputs, implementing agroecological practices. The first vision is compatible with economies of scale, to increase corporate profitability gains for competitiveness in the conventional market and sometimes permitting speculation with capital. The development of the second one is more critical, since it often depends on other resources (NGOs, public support from the Ministry of Agrarian Development in Brazil, research and development), and faces lock-in mechanisms in which the dominant mainstream technology excludes competing technologies (Vanloqueren and Baret, 2009).

The efficiency of agroecology is also under consideration, beyond the presentation of scientific proposals and farmers' experiences showing the potential of agroecology (Agrisud International, 2010; Altieri, 1987; Gliessman, 2007; Pretty, 2006). Assessments mainly focus on one outcome (yields, technico-economic results), whereas agroecology is multidimensional. It is sometimes difficult to identify agroecological specificities: in the absence of formal certification, agroecological practices and situations are often defined on the basis of experts' knowledge. Apart from definition issues (Wezel et al., 2009) and from the tremendous increase in curricula, this raises the question of professionalization in agroecology, especially in Latin America with the expansion in the number of trainees (Sarandon and Flores, 2010). Scientific, practical and societal fields are interconnected in this respect. The challenge will depend on the role of states in the formulation of public policies that can reverse the situation of rural poverty. These entail changes in the economic agendas of governments. In the current climate of crisis and uncertainty, the commitment of individual governments is unclear, although some strategies have served an important number of family farmers. In the Brazilian case, around 1.7 million smallholders of a population of 4.1 million are still marginalized in public policies and live in absolute insecurity. The debate on the future of family farming is still open, albeit polarized around (i) the issue of specialization or diversification in agribusiness and (ii) rural development based on ecological production models, sometimes difficult to implement because the interests and values of organized groups differ.

Some have argued that the dominant food regime has responded to this challenge of food security by promoting a 'weak' ecological modernization process of agriculture, which may mitigate environmental effects to a certain extent, but also cause new negative side-effects and expose some important social, cultural, political and spatial missing links. It may be noted for example that its consequences and implications are the standardization and hygienic bureaucratization of agriculture, the distancing of food production and consumption geographically and institutionally, and the marginalization and fragmentation of the role of agriculture in local communities.

The authors also addressed whether there is evidence in practice that agroecological approaches can contribute to meeting future food demands, especially in developing countries. The arguments are that agricultural practices show a rich variety of multifunctional approaches, mostly implemented by small-scale farmers. On different continents we see examples of farming systems that are locally and ecologically embedded in communities, more resilient towards external threats and globalization, environmentally friendly, contribute to biodiversity and, not the least important, enhance productivity. This should not simply be interpreted as a plea for supporting smallholder agriculture. We have also argued that agroecological initiatives can contribute to food security – which is an urgent challenge for the coming decades – and a value place-based eco-economy. This requires, however, not only analyzing and favouring these initiatives as alternatives to weak ecological modernization, but also a political reorganization of current markets and institutions in ways which reduce the barriers to their wider dissemination and agglomeration outlined above. The real ecological modernization can be up-scaled; but this depends on three major conditions. First, we have to turn the 'problem' of diversity and context dependency of agricultural practices into a real ecological and social virtue. The second condition is an enabling policy (IAASTD, 2008), also acknowledging a new set of counter-logics for the current food regime in, for example, NGO initiatives, some FAO reports and in new institutions like Global Gap and Fairtrade networks. The third challenge is the redirection of agricultural research, development and knowledge transfer. Agroecological approaches could help to 'feed the world' sustainably, and thereby contribute to a 'real green revolution'; but this requires a more radical move and debate among scientists about fostering a new type of (multi-scalar) agri-food eco-economy. This includes rethinking market mechanisms and organizations and more innovative institutional flexibility at different spatial scales, interwoven with active farmers' and consumers' participation. This should be combined with a redirection of science investments to take account of translating often isolated cases of good practice into mainstream agri-food movements. As a matter of urgency it is timely to rethink and critically debate these important issues. We have to question the dominance of the agri-industrial/bio-economy and its particular scientific base as to why more and more people are likely to go hungry while at the same time resource depletion and climate change threaten the Earth's growing population.

Conclusion

The agroecology programme certainly applies to the family agriculture universe, in which the family plays a fundamental role in carrying out and managing agroecological activities. Agroecology's dynamics relies on social movements and approaches rural development issues through discussions and debate in socio-technical networks. As a consequence, its proposition and the diversity of social, ecologically inspired production forms are legitimated, thus modifying the contemporary agrarian scenario.

Agroecology has provoked paradigm shifts and contributed to social criticism of conventional agriculture and, consequently, to the defence of a set of social values linked to the notion of a fair and egalitarian society. Agroecology is legitimated in Latin America thanks to its strong connection with social movements. The agroecology theme brings into public debate the power of science over the progress of society, stressing the political nature of what is behind the technological options underlying the various agricultural systems used by conventional agriculture. Therefore, agroecology may come to play a crucial, determinant role as a catalyser of social criticisms of the importance of the society–science relationship in the contemporary rural world. This is evidenced by the role played by science in agroecology consolidation.

Essentially, the meaning of the movement in defence of agroecology is reflected in the current, manifold rural development process that has recomposed the rural world by restoring the landscape, preserving natural resources, revaluing knowledge and flavour relating to food production, as well as spurring fibre and articraft production and ecological fashion, using renewable energy and re-establishing ethical and human consciousness in the rural world. However statistics are poor, which prevents the advance of socio-economic studies; it is therefore necessary that local institutions make an effort to gather and quantify data corresponding to this process of development. The last national congress in Latin America (2009) had 4,000 people participating, including technicians, researchers, students, representatives of social organizations and small farmers.

From the sociological point of view it is concluded that agroecology is the result of a cultural disruption and transition process undergone by scientific communities and groups concerned about rural development. In practice, the criticism of a production–intensive model translates into a search for agricultural practices of an ecological nature and introduction of alternative innovations such as ecological soil handling and application of social agroecological principles. This is also expressed in the search for alternative distribution methods and fair prices, among other practices and experiments known through the specialized literature. The critical attitude is inherent in social contexts full of latent risks, as can be seen in the agrarian landscape today, and is connected with the global crisis and the future of humanity on Earth (contamination of food, water, air pollution, deforestation, climate change, poverty, hunger, etc.).

This context is intensified by the regulation of production underway in the Latino American continent. This, in turn, makes agroecology a discipline of reference from the standpoint of agricultural practices and search for social equality, being an example for other types of agriculture. Our analysis indicates that the agroecology evolution depends on how strong the interaction of social movements, scientific networks and public policies is, as shown in the case of Brazil.

However, there has been little advancement in the approach to agroecology as applied to agri-food systems. In other words, the matters relating to transformation, conservation and the various kinds of producer–consumer relations are given little prominence in governmental programmes and actions.

Especially in situations of rural poverty or a low level of resources for investment in production facilities may reflect the level of application of the principles of agroecology. Thus, agroecological propositions seem to be in the current context (insufficient funding and policies focused on organized social groups) a matter of survival for poor family farmers and livelihood sustainability.

Notes

1　For further information: 'A produção técnico-científica em Agroecologia' (www.slideshare.net/paventurier/capes-cofecu-bapresentaao27dejunhovdef), whose authors identified over 6,000 references to the subject in Brazil.
2　Recently there have been changes in the family production legislation in Brazil, after acceptance of multiple activities (up to two family members are allowed to work outside the family production unit [PU]). However, the most important income share is derived from the PU, and the PU has to be managed by the family. The farm structure is limited to four fiscal modules (measuring from 20 to 400 ha) according to the municipality to which they belong.
3　The journal *Leisa revista de agroecologia*, June 2011, vol. 27 (2) is dedicated to the topic 'Trees and agriculture'.
4　http://www.clades.cl/publica/publica_index.htm
5　ISEC, Instituto de Sociologia y Estudios Campesinos, was created in 1978. ISEC's empirical work evolves from a formation program that articulates social sciences and participative, interdisciplinary methodologies in a single program.
6　http://agroeco.org/socla/socla_in_english.htm
7　Meeting organized by the 'Associação Nacional de Productres Orgânicos do Peru' – ANPE (National Association of Organic Producers from Peru) at the Universidade Nacional Agrária La Molina – UNALM in 2010. Hundreds of Latin American ecologically inspired producers were present and so were representatives of NGOs and other Latin American institutions and European countries. Among the representatives of organizations , the following stand out: Jannet Villanueva, da ANPE-Peru/IFOAM, Laércio Meirelles, from the Brazilian Eco vida network, Carmen Sotomayor from APOEB, Bolivia, Mauren Lizano, CEDECO/MAOCO, Costa Rica (www.centroecologico.org.br/); (www.aopeb.org/); (www.cedeco.or.cr).
8　Fonte: http://www.planetaorganico.com.br/entrev-galci.htm (Relatório, citado, viagem Lisa Tassi, dissertation Peru, 2010).

References

Abreu, L.S. de (2006) A construção da relação social com o meio ambiente entre agricultores familiares da Mata Atlântica brasileira. Jaguariúna: (SP). Editora IMOPI

Abreu, L.S. de and Watanabe, M.A. (2008) 'Agroforestry systems and food security among smallholder farmers of the Brazilian Amazon: a strategy for environmental global crisis', in ISOFAR Scientific Conference: Cultivating the Future Based on Science, 18–20 June, Modena, Italy, pp. 472–475

Abreu, L.S. de, Lamine, C. and Bellon S. (2009a) 'Trajetórias da agroecologia no Brazil: entre movimentos sociais, redes científicas e políticas públicas', in *Congresso Brasileiro Agroecologia, 6. Congresso Latinoamericano de Agroecologia*, February, Curitiba. Proceedings. Associação Brasileira de Agroecologia, 2, pp. 1611–1615

Abreu, L.S. de, Kledal, P., Pettan, K., Rabello, F. and Mendes, S.C. (2009b) 'Desenvolvimento e situação atual da agricultura de base ecológica no Brasil e no Estado de São Paulo'. *Cadernos de Ciência & Tecnologia, Brasília*, Brasilia, DF, 26, 1/3, pp. 149–178

Agrisud International (2010) 'L'agroécologie en pratiques: guide', Agrisud International, available at: www.agrisud.org/index.php?option=com_content&task=view&id=197&Itemid=277&lang=fr (acessed: 27 August 2011)

Almeida, J. (1988) 'Significados sociais, desafios e potencialidades da agroecologia', in A.D.D. Ferreira and A. Brandenburg (eds) *Para pensar outra agricultura*. Curitiba: Editora UFPR, pp. 277–286

Almeida, G.F. and Abreu, L.S. de (2009) 'Estratégias produtivas e aplicação de princípios da agroecologia: o caso dos agricultores familiares de base ecológica da cooperativa dos agropecuaristas solidários de Itápolis – COAGROSOL'. *Revista de Economia Agrícola*, 56, 1, pp. 37–53

Almeida, S.G, Petersen, P. and Cordeiro, A. (2001) 'Crise socioambiental e conversão ecológica da agricultura brasileira: subsídios à formulação de diretrizes ambientais para o desenvolvimento agrícola'. Rio de Janeiro: AS-PTA

Altieri, M. (1987) *Agroecology: the scientific basis of alternative agriculture*. Boulder, Westview Press

Altieri, M. (2002) 'Agroecology: the science of natural resource management for poor farmers in marginal environments', *Agriculture, Ecosystems & Environment*, 93, pp. 1–24

Altieri, M. and Nicholls, C. (2003) 'Agroecology : rescuing organic agriculture from a specialized model of production and distribution', *Ecology and Farming*, 36, pp. 24–26

Altieri, M. and Toledo, V. (2011) 'A revolução agroecológica na América Latina: esgatando a natureza, assegurando a soberania alimentar e capacitando camponeses', *Journal of Peasant Studies*, 38, 3, pp. 587–612

Beauval, V. (2010) 'L'agroécologie, fertilisant naturel de l'agriculture paysanne?', Campagnes solidaires: Dossier, Bagnolet, no. 251, pp. iv–v

Bellon, S. and Abreu, L.S. de (2006) 'Rural social development: small-scale horticulture in São Paulo, Brazil', in G. Holt and M. Reed (eds), *Sociological perspectives of organic agriculture: from pioneer to policy*. Wallingford, CABI, pp. 243–259

Bellon, S., Lamine, C., Ollivie, G. and Abreu, L.S. de (2011) 'The relationships between organic farming and agroecology', in ISOFAR Scientific Conference, 3, IFOAM, 17, 28 Sep.–1 Oct., Gyeonggi Paldang, Republic of Korea. Proceedings

Bertoncello, B. and Bellon, S. (2008) 'Construction and implementation of an organic agriculture legislation: the Brazilian case', in IFOAM Organic, World Congress, 16, Modena, Italy. Proceedings

Besson, Y. (2011) *Les fondateurs de l'agriculture biologique,* Paris: Editions Sang de la Terre

Brandenburg, A. (2002) 'Movimento agroecológico: trajetória, contradições e perspectivas', *Desenvolvimento e Meio Ambiente*, 6, pp. 11–28

Brasil (2003) Lei no. 10.831, de 23 de dezembro de 2003, available at http://www.planetaorganico.com.br (acessed 02 July 2008)

Buttel, F.H. (2003) 'Envisioning the future development of farming in the USA:

244 L.S. de Abreu and Stéphane Bellon

agroecology between extinction and multifunctionality. Online: http://www.agro ecology.wisc.edu/downloads/buttel.pdf

Caporal, F.R. (2006) 'Documenting agroecology: a competition in Brazil'. Leisa Magazine 22.1, Netherlands, pp. 48–49

Caporal, F.R. and Costabeber, J.A. (2004) 'Agroecologia: aproximando conceitos com a noção de sustentabilidade', in A. Ruscheinsky (ed.), *Sustentabilidade: Uma Paixão em Movimento*. Porto Alegre, Sulina

Dalgaard, T., Hutchings, N. and Porter, J. (2003) 'Agroecology, scaling and interdisciplinarity'. *Agriculture, Ecosystems & Environment*, 100, pp. 39–51

Darnhofer, I., Lindenthal, T., Bartel-Kratochvil, R. and Zollitsch, W. (2010) 'Conventionalisation of organic farming practices: from structural criteria towards an assessment based on organic principles. A review', *Agronomy for Sustainable Development*, 30, pp. 67–81

De Schutter, O. (2010) Report submitted by the Special Rapporteur on the Right to Food, United Nations, General Assembly, Geneva

Francis, C., Lieblein, G., Gliessman, S., Breland, T. A., Creamer, N., Harwood, R., et al. (2003) 'Agroecology: the ecology of food systems', *Journal of Sustainable Agriculture*, 22, 3, pp. 99–118

Gliessman, S.R. (2007) 'Agroecology: the ecology of sustainable food systems', 2nd edn. Boca Raton: CRC Press

Guthman, J. (2000) 'Raising organic: an agro-ecological assessment of grower practices in California'. *Agriculture and Human Values*, 17, 3, pp. 257–266

Guzman, E.S. (2002) 'A perspectiva sociológica em agroecologia: uma sistematização de seus métodos e técnicas'. *Agroecologia e desenvolvimento rural sustentável*, 3, 1, pp. 18–28

Hill, S.B. and MacRae, R. (1995) 'Conceptual frameworks for the transition from conventional to sustainable agriculture'. *Journal of Sustainable Agriculture*, 7, pp. 81–87

IAASTD (2008) *International Assessment of Agricultural Knowledge, Science and Technology for Development: Global Report*, UNDP, FAO, UNEP, UNESCO, The World Bank, WHO, Global Environment Facility, Washington DC

Karam, K.F. (2001) 'Agricultura orgânica: estratégia para uma nova ruralidade'. Tese de Doutorado–Universidade Federal do Paraná, Curitiba

Lamarche, H. (Coord.) (1993) *A agricultura familiar: comparação internacional*, Campinas, Editora da UNICAMP

Luzzi, N. (2007) 'O debate agroecológico no Brazil, construção a partir de diferentes atores sociais', Tese Doctoral thesis – Instituto de Ciências Humanas e Sociais da UFRRJ, Rio de Janeiro

Oelofse, M., Jensen, H.H., Abreu, L.S. de, Almeida, G.F., El-Araby, A., Yui, Q.Y., et al. (2011) 'Organic farm conventionalisation and farmer practices in China, Brazil and Egypt'. *Agronomy for Sustainable Development*, 31, pp. 689–698

Ollivier, G. and Bellon, S. (2011) 'Dynamiques des agricultures écologisées dans les communautés scientifiques internationales: une rupture paradigmatique à rebondissements', in Ecologisaton des Politiques Publiques et des Pratiques Agricoles, 16–18 Mar. L'Isle sur Sorgues, França, available at https//colloque.inra.fr/ecologisation_avignon (acessed 27 August 2011)

Peterson, P. (ed) (2007) 'Construção do conhecimento: novos papeis e novas identidades', in *Uso e Conservafão da Bioversidade*. Cadernos do II Encontro Nacional de Agroecologia. Articulação Nacional de Agreocologia. Recife. Pe, Brasil.

Pretty, J. (2006) 'Agroecological approaches to agricultural development', Background paper for the World Development Report, World Bank

Rosset, P. and Altieri, M.A. (1997) 'Agroecology versus input substitution: a fundamental contradiction of sustainable agriculture', *Society and Natural Resources*, 10, 3, pp. 283–295

Ruis-Rosado, O. (2006) 'Agroecologia: una disciplina que tienda a la transdisciplina', *Interciencia*, 31, 2, pp. 140–145

Sarandon, S. and Flores, C. (2010) 'Teaching teachers: agroecology in Argentina', *Farming Matters*, Buenos Aires, Dec. pp. 36–38

Sauget, N. (1993) 'Une approche américaine de l'agro-écologie', *Natures Sciences Sociétés*, 1, 4, pp. 353–361

Sixel, B.T. (2003) *Biodinâmica e agricultura*, Botucatu: Associação Brasileira de Agricultura. Biodinâmica

SOCLA (2011) http://agroeco.org/socla/documentos_claves.html

Tassi, M.E. (2011) 'Certificação participativa e compra coletiva de alimentos ecológicos: redes locais construindo mercados cooperativos, um estudo na região de Campinas, São Paulo'. Dissertação (Mestrado em Agroecologia e Desenvolvimento Rural) – Centro de Ciências Agrárias, UFSCar. Araras

Tello, J. (2011) (ed) *Agricultura familiar agroecológica campesina en la comunidad andina: una opción para mejorar la seguridad alimentaria y conservar la biodiversidad*, Secretaria da Comunidade Andina, Lima, Peru

Tichit. M., Bellon, S. and Deconchat, M. (2009) 'Agroécologie pour láction', introductory position paper for the General Assembly of the INRA-SAD Division, 27–29 January, Ronéo, SAD, Inra Paris.

Tomich, T.P., Brodt, S., Ferris, H., Galt, R., Horwath, W.R., Kebreab, E., et al. (2011) 'Agroecology: a review from a global-change perspective', *Annual Review of Environment and Resources*, 36, pp. 193–222

Vanloqueren, G. and Baret, P.V. (2009) 'How agricultural research systems shape a technological regime that develop[s] genetic engineering but lock[s] out agroecological innovations', *Research Policy*, 11, 4, pp. 393–412

Venturelli, P., Alencar, M.C F., Ollivier, G. and Bellon, S. (2011) Literatura técnico científica em agroecologia no Brasil: modelo e teste de um método bibliográfico. Manuscript

Warner, K.D. (2007) *Agroecology in action: extending alternative agriculture through social networks*, Cambridge, MA: MIT Press

Wezel, A. and Soldat, V. (2009) 'A quantitative and qualitative historical analysis of the scientific discipline of agroecology', *International Journal of Agricultural Sustainability*, 7, 1, pp. 3–18

Wezel, A., Bellon, S., Doré, T., Francis, C., Vallod, D. and David, C. (2009) 'Agroecology as a science, a movement and a practice: a review', *Agronomy for Sustainable Development*, 29, 4, pp. 503–515

Willer, H. and Yussefi, M. (2008) *The world of organic agriculture: statistics and emerging trends*, in IFOAM Organic World Congress, Modena, Italy, *Anais*

11 Research needs for development of organic agriculture in Sub-Saharan Africa

Charles Ssekyewa, Francisca George and Adrian Muller

Introduction

The world's population was about 7 billion in 2010 and is expected to grow much more. The expected growth is highest in parts of the world that are vulnerable to hunger and adverse climatic conditions, particularly in Sub-Saharan Africa (SSA). This then translates into the likely need for more food in these regions. The situation is exacerbated by the lack of appropriate farming technologies and the many constraints limiting production and productivity of major commodities. Apart from hunger, unless the situation is addressed through a holistic approach to research, there is a very real danger that environmental degradation will escalate.

Nelleman et al. (2009) found that the food crisis in 2007/2008 could even exacerbate in the next decade if there are no explicit answers to the growing new problems, such as declining agricultural production, faltering distribution network and worldwide environmental deterioration. They concluded that food production, processing and consumption across the globe needs to change and that such changes can both lead to food security for the world's rising population and assure the environmental services that are the foundation of agricultural production.

Organic agriculture is one of the most promising approaches to this situation (IAASTD, 2008; FAO, 2008; ORCA, 2009). In addition, organic agriculture is a promising option with regard to further aspects, such as climate change adaptation and mitigation (see Chapter 5). Unfortunately, research funds for organic agriculture are still very low, and only very limited research is going on to enhance the best practices used in organic agriculture.

This chapter discusses the research needs to support organic agriculture as a livelihood strategy, thereby using Sub-Saharan Africa as an illustration. Sub-Saharan Africa is the region with the biggest lack of direct research involvement, financing and infrastructure for organic agriculture in the world. While we will partly focus on those aspects relevant for this region, there are clearly many important aspects that are of relevance for organic agriculture research on a global level.

Research in the context of this book has the main aim to improve livelihoods – not to provide fundamental insights on the world. Linking this discussion to the livelihood framework presented in Chapter 1, and accounting for the particular characteristics of agriculture, we identify five broad areas into which research activities can be grouped. First, there is the *physical basis*: how best to grow various crops and rear livestock organically. In addition, one may ask how best to react to external changes, such as increasing water scarcity. This physical basis has to be understood in a broad sense, also covering the questions related to the farm household, community and landscape seen as sustainably managed systems, decidedly going beyond the level of single crops and practices and their costs and returns for the single household. This point in particular is also stressed by the International Assessment of Agricultural Knowledge, Science and Technology for Development (IAASTD, 2008) report, which talks about moving from focusing on increased productivity alone to holistic integration of natural resource management with food and nutritional security. Second, there is the *social basis*, which has a farming household focus but takes into account aspects of all the different realities, spaces and capitals beyond the physical basis, such as emotional and cultural aspects, the relation of the individual to the community, etc. (see Chapter 1). Third, there is the *institutional context*. This addresses questions such as which policies would be optimal for the support of organic agriculture, which certification approach is most adequate in a certain situation, and also information provision, education and communication. This includes a focus on how optimally to communicate research findings such that they are actually implemented at the farm level, for example. Fourth, *trade and sale* of the products is important. This also addresses how optimally to process and store the produce. It also covers consumer acceptance of new, pest-resistant varieties, for example. Fifth, a particular focus needs to be put on *power relations and inequalities*, e.g. in the context of various certification schemes and value-chain types (see Chapters 6 and 7), but also on a within-household level, e.g. regarding gender issues and how income is distributed in the household.

While discussing these research needs, we do not want to merely report or copy the results from earlier initiatives to compile research needs, such as TP Organics (2009, 2011) or ORCA (2009). Those documents are very broad, highly valuable and should be consulted. There are nevertheless some topics that are not addressed at all or that we propose to address differently. It is those topics we are most interested in. This chapter therefore complements existing work.

In the next section, we briefly present research networks, initiatives and institutions of organic agriculture already present and active in SSA or of general relevance also for this region. We then assess research needs and offer a synthesis of this assessment in the concluding section.

Organic research networks, initiatives and institutions

Organic agriculture research networks, initiatives and institutions have been established globally, regionally and on a national level. These include the International Society of Organic Agriculture Research (ISOFAR) (established in 2003) and the Organic Research Centres Alliance (ORCA) (still in the building phase) on a global level; the Coordination of European Transnational Research in Organic Food and Farming (CORE) Organic I and II (2004–2007 and 2010–2013) and the Technology Platform TP Organics (since 2008) in Europe; the Network for Organic Agriculture Research in Africa (NORA) (since 2009), the West African Network on Organic Agriculture Research and Training (WANOART) in Africa as well as several national initiatives and institutes such as Martyrs University in Uganda, the University of Ghana and the Olusegun Obasanjo Centre for Organic Research and Development (OOCORD in Nigeria). Non-African institutions and organizations that are conducting scientific research on organic agriculture in Africa include the Agro Eco Louis Bolk Institute, the University of Ghent Laboratory of Tropical and Subtropical Agronomy and Ethnobotany (Belgium), University of Kassel (Germany), the Research Institute of Organic Agriculture FiBL (Switzerland) and the International Centre for Research in Organic Food Systems (ICROFS) (Denmark). The web-platform Organic Research (www.organic-research.org) collects information on these and many more networks, initiatives and institutions, and Organic Eprints (www.orgprints.org) is an open-access archive containing papers and project descriptions related to research in organic agriculture.

Furthermore, there is much research going on in contexts that are not specifically related to organic agriculture, but whose results are of great relevance for organic agriculture as well. This is the case for the Association for Strengthening Agricultural Research in Eastern, Central, and Southern Africa (ASARECA), for example. Also the Consultative Group on International Agricultural Research (CGIAR) centres and other research institutions with activities in SSA, such as the World Agroforestry Centre (ICRAF) or the International Centre of Insect Physiology and Ecology (ICIPE) in Nairobi, have many interesting activities relevant to OA. The ORCA database currently (as of November 2011) lists 43 institutions in SSA of potential relevance for research in organic agriculture (www.fao.org/organicag/oa-portal/orca-data base/list/en/).

The most active research bodies appear to be in Nigeria, where several universities are engaged in projects focused on organic agriculture and are involved in partnerships with other African and non-African bodies. The Organic Agriculture Projects in Tertiary Institutions in Nigeria (OAPTIN), based at Nigeria's University of Agriculture, Abeokuta, is engaged in both capacity building and research. The group's efforts are supported by a partnership with Coventry University in the UK. In addition, the National Horticultural Research Institute (NIHORT) has partnered with the University of Ibadan and Ladoke Akintola University of Technology to conduct research, and other work is being done at Ambrose Alli University and two different forestry research centres.

In Uganda, organic agriculture research is taking place at Martyrs University, Makerere University and at the Agro Eco Louis Bolk Institute. Martyrs University is involved in validating and enhancing indigenous knowledge applied by organic farmers to increase soil fertility and improve pest management. Martyrs University is also working to connect with other research institutions to expand its programme in organics. Makerere University is working in research partnership with the International Center of Tropical Agriculture (CIAT) and previously collaborated with the University of Natural Resources and Applied Life Sciences (BOKU) in Austria on the social implications of certified and non-certified organic agriculture (in the 'Linking Farmers to Markets' initiative). Some socio-economic research was conducted in a project funded by the Export Promotion of Organic Products from Africa (EPOPA), though this was mainly a development programme.

Currently there is a research programme, Productivity and Growth in Organic Value Chains (ProGrOV), centred around nine PhD students at three East African universities (Makere, Nairobi and SUA, Tanzania) in collaboration with universities in Aarhus and Copenhagen (Denmark) and coordinated by ICROFS. The ProGrOV programme is taking a value-chain approach and seeks to link research in agro-ecological methods with socio-economic and value-chain research in the companies and market chains of organic products (Box 11.1).

Box 11.1 The ProGrOV research programme in organic value chains in East Africa, 2011–14

Improving productivity and growth in organic value chains in Uganda, Kenya and Tanzania – this is what the research programme ProGrOV is about, by way of developing agro-ecological methods in combination with methods for improved governance and management of value chains. Moreover, development of capacity in research focused on organic and interdisciplinary approaches is an important ambition of ProGrOV. The programme, which builds on collaboration between universities in Uganda, Kenya and Tanzania, is funded by the Danish Ministry of Foreign Affairs and coordinated by the International Centre for Research in Organic Food Systems (ICROFS).

The objective is:

> Research-based knowledge for supporting increased productivity and sustainable growth in organic production and value chains strengthened, and capacity built for future development of the OA-based value chain in Kenya, Uganda and Tanzania.

The rationale and idea:

> There is a need to help smallholder farmers improve their food security by intensification based on improved natural resource management. Organic agriculture is an example of such a strategy where the farmers' improved access to high-value chains and technical input from companies becomes a driver for intensification.

However, the degree and type of improvement in natural resource management and livelihoods for smallholder farmers varies between different organic value chains. Thus, the actual development outcome depends on the dynamics and processes in the product chains. Important factors for this are the character of power relations, the importance of training, differences in the approach to cash crops versus whole-farm development, and the coordination and managerial skills of the intermediaries involved.

Previous projects have demonstrated positive results from development of organic value chains in terms of improved agricultural production. However, they also concluded that chain actors face a number of challenges, such as limited capacity of local farming communities to respond to requirements – including the need to strengthen organic agricultural system development and agro-ecological practices among the farmers.

Therefore, there is a need for integrated research into – on the one hand – how to organize organic high-value chains to improve chain management and livelihood benefits for the farmers and – on the other hand – further developing agro-ecological methods for farming systems intensification based on sustainable natural resource management. While some research has focused on improving productivity and natural resource management of smallholder farmers in Eastern Africa, this has in most cases not been associated with studies of how to link the improved production output to market access and quality demands. This is why the ProGrOV project attempts to integrate research along selected organic value chains in a cross-disciplinary approach, combining a number of PhD studies covering organic pest management and crop nutrition for improved yield and product quality, integration of livestock in organic cash crop farms, and chain management for the export market, high-end domestic market and the tourism sector (Figure 11.1).

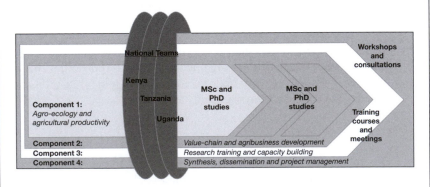

Figure 11.1 The organization of the value-chain-focused research programme, ProGrOV, demonstrating the interlinkages of agro-ecological and agribusiness projects

ProGrOV project approach

In the ProGrOV programme production elements of organic value chains such as pineapples and perishable vegetables will be studied with respect to how farmers can best respond to and collaborate with the demands of the intermediaries and markets. Collaboration with the chain actors, such as farmers, processers and traders, is essential for all the PhD studies. Therefore, the value-chains focus of the project is combined with participatory approaches, in order to ensure that the knowledge generated through the studies will be useful and applicable to real-life situations.

This is illustrated in Figure 11.2. The upper side of the diagram represents the research process, which is informed by the stakeholders, i.e. the national organic organizations, farmers, private companies and selected markets such as local supermarkets, etc. The research questions and research findings are tested in value-chain stakeholder forums. These thus act as a dissemination forum for a reality check for the researcher, as well as a forum where fine-tuning and adjusting the research can take place. The fora obviously differ along the value chain. Thus, if one assumes that a certain input of livestock manure could improve the amount and quality of vegetables, then before testing this intervention experimentally it is necessary to discuss the feasibility of the intervention with the farmers (i.e. would they potentially be able to find and use equivalent amounts of manure?). The product quality in the ProGrOV project is a relative and context-dependent concept which will also be guiding the research as described by Høgh-Jensen (2011).

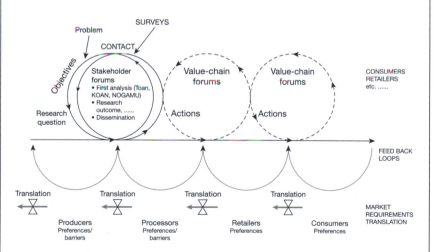

Figure 11.2 Schematic representation of innovation research in primary value chains involving relevant stakeholders

Capacity building is a strong element in the project. The project is based on nine PhD studies and six MSc studies at the three universities in Tanzania, Uganda and Kenya. The project organizes special training courses in value-chain, participatory and multidisciplinary approaches. The students will implement the studies in close collaboration with the national organic organizations in the three countries (TOAM, NOGAMU and KOAN). The knowledge developed in the studies will be synthesized towards the end of the project and disseminated to key stakeholders.

The Agro Eco Louis Bolk Institute is currently developing an organic agriculture research and dissemination network with Martyrs University, the University of Ghana, the University of Ibadan (Nigeria), the Institute of Insect Physiology and Ecology (ICIPE, Kenya), the World Agroforestry Centre (Kenya) and the Rainman Landcare Foundation (Zambia).

Leading organic agriculture research institutions in Kenya include the Kenya Agricultural Research Institute (KARI), ICIPE and the Kenya Institute of Organic Farming (KIOF) as well as Egerton University. A seven-year research trial with organic farming strategies and crop rotations was set up in collaboration with BOKU and supplemented with case studies among farmers. KARI often offers in-kind support and some funding to groups working on organic agriculture. In Kenya, the University of Munich was involved in research at Malindi to assess the sustainability of organic agriculture by using various indicators. Maseno University conducts research on indigenous vegetables. Furthermore, FiBL is involved in conducting research on systems comparison trials in the tropics and access to markets, as well as waste management in cities. The systems comparisons in the tropics are located in Kenya, India and Bolivia. These long-term farming systems comparison field trials were established in 2007 and aim at comparing the performance of conventional and organic agricultural production systems in the tropics. They are guided by FiBL, and the systems comparison trials are complemented with participatory on-farm research.

Ghana has an ongoing cocoa research project. Here the focus is on the black pod disease and capsid pests, as well as on improving soil fertility. In Mauritius, the Agricultural Extension Unit (AREU) partners with government agencies and the University of Mauritius to conduct research.

Topically, much of the research being conducted to date on organic agriculture in Africa is at the crop and management practice level and is concerned with the impact of compost and soil amendments on crop yields, the identification and evaluation of the efficacy of bio-pesticides, and the use of legumes for nitrogen fixation. It is interesting to note that there is not much research conducted on organic livestock production in Africa.

It should also be mentioned that some of the research carried out by established organizations in agro-ecological methods such as soil fertility management, integrated natural resource management and non-chemical pest management is relevant to organic agriculture. The ASARECA includes in its strategies topics which are relevant for organic agriculture such as *Promoting utilization of integrated soil fertility management technologies for major food and high value crops*; *Management of problematic soils for improving agricultural productivity*; and thematic areas focusing on *Policy, institutions and governance for sustainable natural resource management* and on *conservation and management of agro-biodiversity for wider use* (ASARECA, 2009). For example in the project 'Building sustainable rural livelihoods through integrated agro-biodiversity conservation in ECA savannah ecosystem' one of ten 'best bet practices' to be tested in the area is organic farming (www.asareca.org/researchdir/PROGRAMME_DONORB006.HTM).

Moreover, ICRAF develops integrated systems, so-called Evergreen Agriculture, based on organic inputs and nutrient recycling from nitrogen fixating trees to annual crops which could be very important for the up-scaling of organic agriculture (Garrity et al., 2010).

Another example is that ICIPE has developed the so-called 'Vuta sukuma – a pull-push system for stem borer and striga control for maize and sorghum in Africa' based on the development of in-depth scientific knowledge on pest and predator behaviour, host reactions and allelopathic effects as well as their interrelationships. The research programme is an interesting combination of molecular and chemical laboratory science, controlled experiments, agronomy and participatory on-farm research and the end-users have been involved throughout long periods of the research in order to ensure usability and uptake (information available at www.push-pull.net/index.shtml).

Although this list could be prolonged and reports an impressive record of organic agriculture research, much more needs to be achieved. Organic agriculture research has not yet found a place in most research institutions, in particular in African contexts. The main reasons for this, for example as identified and agreed on at the African Organic Agriculture Conference 2009, are:

- African research is only just beginning to develop interest in organic agriculture research, development and practice;
- funding for OA research is scarce;
- OA promoters have not created appropriate personal relations with researchers;
- OA is seen by some stakeholders more as a niche trade issue (i.e. marketing of high-value certified cash crops) rather than an alternative farming system, especially by the CGIAR centres research; outputs generated by conventional agriculture researchers are not seen to be relevant to OA farmers' problems;
- OA research is not considered fashionable in modern times;
- there are limited opportunities for publishing OA research findings.

However, it was also observed at this conference that several students were interested in OA research due to existing global environmental concerns. It was therefore proposed that to enhance OA research in Africa, there is a need to:

- establish private OA centres of excellence to conduct research;
- initiate and fund more South-to-South partner networks to eliminate the over dominance by North-based research institutions;
- encourage knowledge exchange among researchers in similar farming conditions;
- establish an organic agriculture journal;
- bring on board as many interested young researchers as possible;
- increase efforts to look for OA research funding from like-minded institutions;
- ensure a participatory approach to research.

The establishment of an Organic Agriculture Network was considered by participants to be timely. It was suggested that linkages be made with some international organic and research organizations, such as the International Federation of Organic Agriculture Movements (IFOAM), the International Society of Organic Agriculture Research (ISOFAR), the Forum for Agricultural Research in Africa (FARA) and the Association for Strengthening Agricultural Research in Eastern and Central Africa (ASARECA). The need for a very effective networking structure was recognized.

Participants at the African Organic Agriculture Conference indicated that they would like to see an OA network for African organic stakeholders that has a research results database in local and regional languages, conducts local and regional face-to-face meetings, recognizes grassroots research, is complementary to existing research and development projects and initiatives and has a two-way flow of research needs and results. Thus, the Network for Organic Agriculture Research in Africa (NOARA) was launched during the conference in 2009, in Kampala, Uganda. However, NOARA is still in the building phase, funds are scarce and much still needs to be done. Although not linked to Africa specifically, the establishment of the journal *Organic Agriculture* in 2011 can be deemed a success as it bears the potential for bundled publication of high-level research on organic agriculture.

Research needs and their assessment

The previous section illustrates that there are a considerable number of institutions involved in projects on organic agricultural research in SSA, addressing topics in all five of the broad areas identified in the introduction (i.e. physical basis; social basis; institutional context; trade and sale; power relations and inequalities). However, the single projects seem not to be integrated in a wider common context and an encompassing research strategy seems lacking. The most recent initiatives that could provide such an overarching view, for example, the HUSHA proposal (ORCA, 2009) or NOARA, are not yet established, or seem not to take off. In this and the next section, we present research needs for organic agriculture, with some focus on SSA. We structure this assessment along two main types of approach – top-down and bottom-up – on how to identify these needs. We then specifically address biotechnology as one controversial field of research in organic agriculture and end with a synthesis of this assessment.

General research needs assessments

There already exist some assessments of research needs in organic agriculture in general (e.g. TP Organics, 2009, 2011) and with relevance for SSA in particular (e.g. ORCA, 2009; NOGAMU, 2006). Key research needs are identified in several broad categories.

For SSA, ORCA (2009) identifies three main challenges for developing organic agriculture: (1) building and maintaining soil fertility, nutrients and water-holding capacity, (2) the problems of pests and diseases (which will become even more important in the context of climate change, where pest and disease pressure is predicted to increase), and (3) optimal structures for capacity, knowledge and skill building of farmers, to enable them to manage the complex organic production system. The ORCA idea suggests an organization in a number of centres based on agro-ecological zoning in combination with a division of responsibilities for key thematic areas of relevance to organic food and farming. Thus, the proposed centre on 'Arid and Semi-Arid Agro-Ecosystems' is foreseen to be leading in the following research area:

> Organic farming methods, which are more resilient to drought conditions, could be the solution against poverty and hunger, reducing overgrazing and improving soil fertility. The main challenge of converting to organic agriculture in this agro-ecosystem is dealing with the scarcity and the disrupted dynamics of biomass decomposition during the long dry season(s) which result in a very slow build-up of soil organic matter. Research on best practices for animal husbandry to raise livestock with high productivity within an arid and semi-arid organic system is another need.
>
> (www.orca-research.org/orca-centre-arid-semi-arid.html?&L=idtbbcdeumlpvldb)

UNEP-UNCTAD (2006) emphasizes the following aspects for East Africa:

> More research needs to be carried out in this area to show how organic agriculture benefits resource poor households, especially in regard to women and children and whether commercializing smallholder farmers really leads to a decrease in poverty, or whether the man of the household is the sole beneficiary of the extra income.

And as a country example, the NOGAMU (2006) research needs assessment identifies four key topics: (1) pest and disease management strategies in organic agriculture; (2) soil fertility improvement strategies; (3) high-yielding varieties suited to organic production systems; (4) intercropping combinations. Of great interest is the IFOAM report (2009) which focuses on women's empowerment in organic agriculture and identifies a range of related research needs.

In preparation for the ProGrOV project (Box 11.1), the Kenyan Organic Agriculture Network (KOAN) identified a number of research needs ranging from classical natural resource-management questions (soil fertility management, water harvesting, organic practices for pest management) to comparison trials with conventional farming to socio-economic and consumer-related topics. Thus, besides the agro-technical topics, KOAN suggested research into:

1   assessing the contribution of organic agriculture to food security
2   assessing the contribution of organic agriculture to the GDP
3   assessing the contribution of organic agriculture in improving livelihoods
4   assessing the contribution of organic agriculture to environmental con-
    servation: biodiversity, mitigating or adapting to climate change (carbon
    sequestration, etc.)
5   assessing and documenting farmers' knowledge and innovations in crop and
    livestock production, human and environmental health
6   profiling the organic producers and consumers
7   assessing the trends and attitudes in the organic marketing chain.

While ProGrOV includes aspects from the last points on consumers and markets, these topics need significantly more research in Africa, including the potential for markets in the tourist industry. The request to include farmers' knowledge in a research agenda is in line with the recommendation in the IAASTD reports (Chapter 1). There is a need for a more thorough research needs assessment for organic agriculture and food systems related to different regions of Africa in order to develop a proper research strategy, which may be used to influence and inspire national and international research funders.

While implementing such research, an 'alternative way of undertaking research' should be followed, claims ORCA (2009):

> Research in organic food production requires a holistic approach of com-
> bining traditional and indigenous knowledge, social development, technical
> innovation and market development. It builds on the following principles:
>
> • Participation of all actors in the food chain, including farmers, pro-
>   cessors, traders, advisers, trainers and policy makers;
> • Networking of institutions in order to mobilize all the expertise
>   required from organic research institutions as well as other research
>   institutions with interest in organic research;
> • Cross and inter-disciplinary research building bridges between dis-
>   ciplines and between science and society;
> • Holistic systems methodologies that can describe the production from
>   a systems perspective;
> • Global collaboration, which will include twinning arrangements
>   between organic research institutions in developed countries and in
>   developing countries.

Finally, we may also mention the assessment of the 'top 100 questions of importance to the future of global agriculture' (Pretty et al., 2010) as a catalogue of research needs. They cover a very broad range of topics, are kept quite general and apply to agriculture in general and not only to organic agriculture. The latter point is important, as many research needs formulated for agriculture in general, especially if they do not relate to single management practices, are also research needs for organic agriculture.

Bottom-up research needs assessments: an example from Uganda

While general assessments as presented in the previous sub-section help identify broad key areas of research need, another route is also being taken, identifying more specific knowledge gaps at farm level. Of these gaps, many could be remedied without further research but by appropriate information provision and extension services, based on a large body of research results that are readily available. Others, however, may indicate some mismatch between state-of-the art research and applicability for the farmers. Therefore research is needed on how to optimize information provision, extension services and implementation of research results. The training manual of Organic Africa launched in 2011 (www.organic-africa.net/oa-home.html) could be a context where this could be undertaken, if the success of its implementation and uptake were analysed in detail. Third, some knowledge gaps are due to genuine lack of research results. More research is needed there.

As an example of such bottom-up research needs, we present some results from the survey from 2006 undertaken in Uganda by NOGAMU in more detail (NOGAMU, 2006). The survey included 223 farmers from all parts of Uganda. To be able to gauge their knowledge on organic farming they were first asked about pertinent issues and practices in organic agriculture. This helped generally to assess their training needs and to differentiate these from genuine research needs. As indicated in Table 11.1, knowledge about organic practices concerning seeds, nursery beds, transplanting and insect pests and disease management was below average among the respondents. On the other hand, knowledge on organic weed management, fertility management and post-harvest management was above average.

Not only was their knowledge about pests and diseases below average, the farmers indicated those as the major limiting factors in their production. A detailed survey on pests and diseases of the most important crops of those organic farmers was then undertaken. This helps identify where appropriate knowledge provision and extension services could remedy the problem, and where genuine research is needed – be it on adequate dissemination and implementation of already known results at the farm level or on areas where genuine knowledge gaps prevail. In this crop-wise assessment, areas for genuine research, for improvement in research application and for extension services on well-established knowledge can readily be identified.

Table 11.1 Knowledge assessment of the participants in the survey

Scores	1	2	3	4	5
Topics	*No knowledge*		*Average*		*Expert*
Seeds	50 (21.7%)	72 (31.3%)	51 (22.2%)	35 (15.2%)	22 (9.6%)
Nursery bed	105 (48%)	48 (21.9%)	34 (15.5%)	20 (9.1%)	12 (5.5%)
Transplanting	79 (36.2%)	38 (17.4%)	57 (26.1%)	26 (12%)	18 (8.3%)
Field establishment	9 (4.5%)	37 (18.7%)	79 (39.9%)	50 (25.3%)	23 (11.6%)
Weed management	10 (4.5%)	48 (21.4%)	94 (42%)	46 (20.5%)	26 (11.6%)
Insect pest management	48 (21.3%)	59 (26.2%)	66 (29.3%)	33 (14.7%)	19 (8.4%)
Disease management	61 (27.5%)	58 (26.1%)	60 (27%)	27 (12.2%)	16 (7.2%)
Fertility management	15 (6.6%)	35 (15.4%)	83 (36.4%)	49 (21.5%)	46 (20.2%)
Post–harvest	26 (12.2%)	36 (16.9%)	65 (30.5%)	56 (26.3%)	30 (14.1%)

Notes
Scores run from 'no knowledge' (score 1) over 'Average' (3) to 'Expert' (5). The numbers indicate the numbers of respondents in each category, percentages in parentheses.

Source: NOGAMU, 2006.

Coffee

Stem borers were found to pose the biggest pest problem in coffee and coffee wilt disease was reported to be the worst disease, both in low and highlands. Some symptoms mentioned relate to mealy bug damage. The yellowing reported requires following up during verification surveys. The farmers did not seem to have a systematic way of managing and controlling pests and diseases in coffee. Farmers' management of pests and diseases in coffee seemed to be based on cultural practices that acted as preventive measures, consisting mainly of cultural practices like pruning, roughing and general field hygiene. Apart from this, control measures were applied after the symptoms were observed to have caused a loss in the harvest. Although different plant concoctions were used for control of pests and diseases, they were not used on a regular basis. Their use was prompted by the severity of attack of a disease or pest.

Cocoa

Thrips were reported to be the biggest pest problem in cocoa production, followed by aphids and weevils in order of importance. Farmers seem, however, to be mixing up pest problems, and it was found that while beetles were reported, for example, the pest could actually be weevils. In addition, white flies were not understood to be a pest of cocoa. There is need therefore to follow up and ascertain some of these pest issues. Black pod and die back were reported as the worst diseases. These could be symptoms of attack by pests or infection by some diseases, all of which require verification. Cocoa farmers did not seem to have any means of controlling pests and diseases on their cocoa fields. No solution was reported as being used to control any of the observed pest and diseases. This could mean that all the observed cocoa pests and diseases were not of economic importance or it could be attributed to total lack of knowledge about potential solutions.

Cotton

Aphids, cotton lygus and boll worms in receding order were noted to be the most important pests of cotton. Farmers used cultural and physical methods of control, as well as biorationals and natural enemies (Nginingini). There was only one disease regarded as of some importance (Fusarium wilt), but it was at low incidence only and this was controlled by enforcing government regulations to use dressed seed. This could mean that diseases had been contained by Ministry of Agriculture, Animal Industry and Fisheries (MAAIF)/CDO (Cotton Development Organisation) policy on acquisition of cotton seed. Although coffee and cocoa had a higher number of mentioned pest problems (nine and seven, respectively), cotton, with only six mentioned pest problems, had the highest number of solutions to pest problems. This may be attributed to the higher level of pest and disease awareness and knowledge observed in those cotton growing areas, which can be due to the economic importance attached to the crop in that area. It could also be that there are natural control means taking place, though. This needs verification.

Pineapple

Mealy bugs were reported to be the commonest and worst pest problem in pineapples. Bacterial heart rot was found to be the worst disease problem. Although symptoms of probable nematode attack were visible, no pineapple farmer reported it. This means that farmers were not aware of the symptoms of nematodes. Farmers did not seem to have many practical solutions to the pest problems reported. Use of Tephrosia extract was the major control against observed pests (except monkeys). The only direct method of disease control was by roughing. This could mean that either all the observed pest and disease problems are not of economic importance or that pineapple farmers are not aware of appropriate control methods. The fact that the association between black ants and mealy bugs was not known and that farmers were not aware of the symptoms of nematodes (which were visible in the field) also points to some lack of knowledge.

Sesame

Out of the 50 sesame farmers interviewed, only 18 reported pest problems. The golden midge was the most frequently reported pest. Only one disease was mentioned, by only one farmer. According to farmers in the northern region, sesame had no major pest problems. The pests and disease that were mentioned were sporadic and were not of major economic importance. This could be attributed to resistance of the variety of sesame grown by the farmers.

Vanilla

Although originally vanilla was said to have no pest problems, farmers have noted the slow emergence of some pests and diseases in vanilla over the past 10 years. Either the farmers did not understand vanilla problems or new vanilla problems are arising due to spread of vanilla out of its original ecosystem. The pest related to black ants needs to be identified. Pests including thrips, aphids, fruit flies and some caterpillars were reported by some farmers as occasional problem. The numbers of pest problems reported in vanilla were as many as those reported in coffee. Thrips were the pests most reported by farmers, while the commonest disease was rust. The solutions applied for vanilla, however, were mostly limited to improving soil fertility to support plant resistance. Vanilla farmers saw organic manure as the basis of management of both pests and diseases.

Banana

There were few pests and disease types reported in banana production, but all of them were of major importance to the farmers. Banana weevils and banana wilt were reported to be the worst constraints. Farmers, however, lacked ample knowledge on disease problems. They could not identify separately the symptoms of disease and symptoms of soil deficiencies. In particular, there is need for urgent training on management of nematode infestation in banana fields. Nematodes were reported once and no solution against them was reported, yet their symptoms were evident. And if symptoms of nematode infestation are detected (toppling of banana after rain), for example, farmers respond by defoliating all the remaining bananas. This practice reduces yields further, yet it does not solve the nematode problem.

Ginger

Although there had been reports of cases of ginger wilt in areas not surveyed during this assessment, according to ginger farmers in the study area, ginger had almost no problems in its production. There were only a few occasional infections of the leaves, but these are of minor importance.

Capsicum

Fruit flies were observed to be the biggest problem in capsicum production and the farmers had no solution. Capsicum is a relatively new crop in organic farming and research is needed to address the fruit fly problem. Viruses were reported as the biggest disease problem. This was often related to seed quality problems due to the continuous recycling of seed practised by farmers. This may not be true, as it could also be attributed to lack of appropriate control of vectors at nursery stage, leading to virus infection. There is need for follow-up to classify the viruses for their proper systematic management. Generally, capsicums were reported as a difficult crop to produce, whether organically or conventionally.

Passion fruit

The pest problems in passion fruits did not seem to be well articulated by the farmers. The most reported problem was that of ants which could be a sign of the presence of aphids or mealy bugs. To the farmers, the ants seemed more harmful than aphids, because the latter hardly moved. The woodiness disease was the commonest one reported by passion fruit farmers. This could be attributed to lack of proper identification and handling of the vectors. The situation regarding pests and diseases in passion fruit production shows that the farmers have not yet mastered the techniques of pest identification and management.

Finally, this survey collected information on the most important areas for research as formulated by the farmers themselves (see Table 11.2). Besides analysing the issues that were of biggest importance (pest and disease management and organic pesticides and herbicides), it is also of interest to look at those that seem not to pose problems at all (e.g. water harvesting and irrigation). Potential research needs on institutional aspects such as certification issues, etc. were not mentioned, but this is likely due to the formulation of the relevant question in the questionnaire, that clearly focused on agricultural practices and techniques.

Table 11.2 Potential areas for research as formulated by the participants in the survey

	Suggested research area	*Crop/product*	*No. of respondents*
1	Pest and disease identification and management including weeds	Cotton, sesame, other crops	125
2	Organic pesticides and herbicides	General	38
3	Market research	General	29
4	Higher-yielding variety/seeds	Cotton, sesame, hot pepper, passion fruit, coffee, other crops	20
5	Soil testing for fertility conservation	Cotton, sesame, other crops	16
6	Improvement of productivity	Cotton, sesame, general	7
7	Product development and processing	General; more on pineapples, fish	6
8	Organic agriculture agronomic practices	General	5
9	Spacing in sesame and cotton	Cotton, sesame, other crops	3
10	Water harvesting and irrigation	General	1
11	Labour substitution	General	1
12	Quality issues in OA	General	–
13	Standards	General	–
14	Certification	General	–
15	Training and capacity building	General	–

Source: NOGAMU, 2006.

Summarizing, we could state that the farmers' emphasis on the importance of pests and diseases may identify research needs in this context. On the other hand, the exemplary study described above shows that some of the pest and disease problems may be symptoms for some other underlying cause, such as poor soil fertility and plant health. Thus, improving extension services on these aspects may solve the perceived problems, and research needs may rather be located in these topics than in pests and diseases. To identify research needs, it is thus crucial to clearly identify the causes behind some actual situations.

Taboos and biotechnology

A particular aspect of research in both organic and conventional agriculture is taboos, rooted in the convictions of what these farming systems must be like or not. In conventional agriculture, an example of a taboo may be research on the concepts and interlinkages of soil fertility, soil health and plant health as a basis for a productive agricultural system. This taboo has been weakened recently, due to mitigation of climate change in agriculture, where soil organic matter is a key aspect. In organic agriculture, a taboo is to do research on practices that somehow are judged as 'unnatural', e.g. much of biotechnology; although some biotechnologies are compatible with organic agriculture, some are even widely used already, and some are promising and thus deserve to be researched objectively. As an illustration, we will deal with biotechnology and related research in organic agriculture in some more detail. This section draws strongly on Ssekyewa and Muwanga (2009).

For this discussion, we adopt a rather broad view of biotechnology and understand it as a range of methods that are used to manipulate living organisms to meet human needs, be these methods ancient or modern, simple or complex. In this understanding, biotechnology covers the application of indigenous and modern scientific knowledge to the management of organisms, organs, tissues, cells or their organelles to supply goods and services to human beings (Bunders et al., 2005). A similarly broad understanding of biotechnology is adopted by the Convention for Biological Diversity, the Cartagena Protocol and also by the IAASTD (2008) report. The organic community in general adopts a narrower understanding of biotechnology, as reflected in IFOAM's position on genetic engineering and genetically modified organisms, where they even see a threat in using the term 'modern biotechnology' as it may only divert from the fact that the main focus of the organic community is GMOs (IFOAM, 2002). They thus focus on genetic engineering (GE) which they define as: 'a set of techniques from molecular biology (such as recombinant DNA) by which the genetic material of plants, animals, micro-organisms, cells and other biological units are altered in ways or with results that could not be obtained by methods of natural mating and reproduction or natural recombination' (IFOAM, 2002). This sort of manipulation may be applied in seed production, animal breeding or for production of inputs.

For organic stakeholders, biotechnology usually embraces genetic engineering or basically coincides with that and thus becomes suspect. Stakeholders from the organic community reject genetically modified materials on such grounds as the following:

- *Unnaturalness*: application of biotechnology (namely, GE) leads to (parts of) organisms that are not naturally occurring, different to the outcomes from other breeding techniques such as hybridization in open pollinated plants.
- *The precautionary principle*: any application of untested technology with unknown effects must be avoided.
- *Health risks*: it is not clear what health risks may be taken by consumption of genetically modified food.
- *Environmental risks*: genetically modified crops can pollinate wild weeds and neighbouring crops. This can have adverse consequences, such as in the case of Bt (bacillus thuringiensis) crops, where damage to non-target insects can occur (Altieri, 2005).
- *Dependency*: genetically modified crops are likely to increase the dependency of farmers on seed companies; take terminator genes as an example.
- *Exclusivity/undermining alternative methods*: widespread introduction of Bt crops, for example, poses a special threat to organic farmers, as it may foreclose their moderate use of Bt as a biological pesticide due to inducing resistance to Bt in many pests.

When discussing biotechnologies, it is important to note that these arguments for rejection refer to very different underlying reasons and embracing one argument does not mean acceptance of others. The unnaturalness argument, basically reflecting the dislike of 'unnatural' interference with generation/reproduction of organisms, is highly normative and will be difficult to argue consistently in modern societies. The precautionary principle and health and environmental risk arguments address the huge uncertainties and potential adverse effects involved and the level of risk a society is ready to take. Part of the environmental and health risk arguments can be strengthened or criticized on the basis of an ever-growing body of established knowledge. The dependency argument refers to questions of equity and justice and has to be seen in the context of the current highly concentrated corporate structure of modern agriculture (cf. Chapter 8), the efforts undertaken by these key players to patent genes, organisms and related technologies, the resources these companies invest for lobbying work and the resources needed for some biotechnologies such as GE. Exclusivity, finally, refers to an efficiency argument. Using an effective substance to such an extent as to support widespread development of resistance among the target pests may be efficient in the short run, but not so in the long run, when this substance has essentially become useless (cf. the general situation with overuse of antibiotics).

While these counter-arguments clearly apply to genetically modified crops, and while genetically modified crops are prohibited in organic agriculture (IBS, 2005), undifferentiated rejection of any biotechnology is no response. It would foreclose the possibility that organic farmers may profit from the new biotechnological innovations that pose no environmental problems and that could be adopted in organic farming without violating any principles or guidelines, and that are partly also widely in use already.

Simple existing (traditional) practices that could be seen as biotechnology are the use of cuttings, manipulated cross-breeding, removal of seeds from shells to break dormancy, artificial insemination, pruning, topping of fruit trees, spraying with plant extract oils, mycorrhiza associations, etc. (Stoll, 1998). These simple practices can be very successful. Using coffee clones from cuttings, for example, is responsible for the sustained coffee industry in Uganda which thus managed to deal with the incidence of coffee wilt disease. Using cuttings and grafting onto resistant rootstocks allowed the control of collar rot disease of passion fruit in the Common Market for Eastern and Southern Africa (COMESA) region, where production increased by over 50 per cent (Ssekyewa et al., 1999). Since colonial times, grafted buds are the base of the citrus industry. The desired quality of fruits is achieved by budding, and the results are achieved in a shorter time than when seeds are used (Hartmann et al., 1990).

Next are more complex techniques such as plant tissue culture for micro-propagation and production of virus-free planting materials and molecular diagnostics of crop and livestock diseases. These may be called first-generation modern biotechnologies. These technologies are easily applied and not costly and thus have already been adopted in many developing countries. Several examples of successful implementation of such technologies exist. One is the production of disease-free sweet potatoes that have been developed via tissue culture and have been planted on 500,000 hectares in Shandong Province, China. This led to yield increases of about 30 per cent (Fuglie et al., 1998). Another example is the global eradication of rinderpest that was supported by advanced biotechnology-based diagnostic tests (Roeder and Rich, 2009). New cassava (*Manihot esculentum*) planting material has been developed from meristem tissue culture. This ended the cassava planting material scarcity in Africa that was due to the cassava mosaic virus (Ogero et al., 2010). In eastern, central, and southern Africa, new bean varieties (*Phaseolus vulgaris*), many with multiple stress resistances, have been developed through marker-assisted breeding and farmer seed multiplication (Witcombe et al., 1998). They play an important role and are grown and consumed by nearly 10 million farmers, mostly women. Further techniques include in vitro production of Solanum potato micro-tubers, bio-prospecting in traditional medicine (Twarog and Kapoor, 2004) or biopesticides that are created by cells of infected tissues and can be extracted using modern techniques.

Much more complex are so-called second-generation modern biotech-nologies that rely on molecular biology techniques and use genetics to provide information on genes important for particular traits. In these techniques, DNA or RNA is extracted, isolated, bulked, and specific gene markers are identified. Through comparison studies, the existence of desired genes in other varieties or animal breeds can be identified. Information generated by each of these activities can then be used in natural breeding programmes. As *The World Development Report 2008* states:

This allows the development of molecular markers to help select improved lines in conventional breeding (called marker-assisted selection). Such markers are 'speeding the breeding,' leading to downy mildew-resistant millet in India; cattle with tolerance to African sleeping sickness; and bacterial leaf blight resistant rice in the Philippines. As the costs of marker assisted selection continues [sic] to fall, it is likely to become a standard part of the plant breeder's toolkit, substantially improving the efficiency of conventional breeding.

(WDR, 2008: 163)

Application of biotechnology to fight pests and diseases that are about to emerge will probably add to the list of examples of successful implementation of biotechnology, e.g. against coffee wilt (*Fusarium* spp), banana wilt (*Xanthomonas* spp.), maize grey leaf spot (*Cercospora zeae-maydis*), woolly white flies on citrus (*Aleurothrixus flocossus*), tomato leaf curl (*Begomovirus*) and others. However, knowledge on and understanding of biotechnology are still very scanty in Sub-Saharan Africa, in particular among farmers and in organic agriculture training (Markwei et al., 2008; Paul and Steinbrecher., 2003).

Synthesis of research needs and conclusion

This and the other chapters in this volume have shown that there are research needs in all five broad topics as identified in the introduction: physical basis; social basis; institutional context; trade and sale; power relations and inequalities. As we pointed out earlier, the following assessment will not be encompassing, and we refer the reader to consult the general research needs assessment as presented in TP Organics (2009) or ORCA (2009) as well. Here, we focus on some complementary key points and aspects that are not or only partially covered in these assessments.

Physical basis

The previous sections have shown that there is already much research being undertaken regarding the physical basis of organic agriculture and more is still needed. Biotechnology not involving genetic engineering has great promise, but much research is needed on technical aspects for the particularities of organic agriculture. For example, suitable organic substrates for tissue cultures should be developed.

Furthermore, much research is still needed on classical aspects of agricultural practices. Research needs are formulated predominantly for the physical basis, e.g. for various pests and diseases and on improved control thereof. Pests are not only an issue on the fields, but they are also responsible for storage losses. Given that in the global South 30–40 per cent of all produce is lost in storage (Godfray et al., 2010), considerably reducing such losses is one of the single most powerful measures to increase sustainability in agriculture. Reducing losses by 15 per cent, for example, would correspondingly reduce the pressure to increase yields, which can be a more challenging task. There are low-hanging fruit to be harvested in this context, but currently only a few institutions research this topic. The Kenya Institute of Organic Farming (KIOF) is an example, and one of the proposed ORCA centres: the Centre on Post Harvest and Safety would specifically address storage losses as part of one of its focuses.

The assessment in previous sections showed, though, that part of the demand for research on the physical basis likely could be met with better information provision and increased quality of extension services. This is much cheaper than more research, and identifying the true research needs thus avoids spending scarce financial and human capital resources where they are not productive.

Correspondingly, there are research needs on aspects of optimal information provision and on why well-researched and promising optimal practices are not implemented in a certain context. There, the focus is no longer the physical basis, but rather on the various livelihood realities, spaces and capitals beyond physical aspects (see Chapter 1), i.e. the *social basis*: most smallholder farmers are risk-evasive and use mostly indigenous practices (Ssekyewa, 2008). To assure optimal organic production systems, stakeholder attitude towards both traditional and modern biotechnology has to change. It has to be widely acknowledged that organic agriculture is not traditional farming and does not negate all modern technologies. Again, to achieve this, extension officers also need to be aware of the potential of modern biotechnologies. Taboos hindering a thorough and critical assessment of all available options need to be overcome. Social science research, for example on acceptance of new technologies and the spreading of innovation, plays a crucial role and a tight combination of training and research is needed. It is also very important to meet the research needs on gender aspects in organic agriculture, in particular on the role and empowerment of women (cf. IFOAM, 2009).

Institutional context

Current investments in biotechnology are concentrated largely in the private sector, driven by commercial interests, and not focused on the needs of the poor, who are the majority in the organic sector in SSA. Increasing public investment in research on pro-poor/smallholder farmer traits and crops is needed and the capacity to evaluate the risks and regulate these technologies needs to be improved. An example illustrating the research needs on an institutional level is: which institutional design is best suited to support organic agriculture? In agricultural policy, payments for environmental services are one most promising approach, and area payments for organic production also perform well, as they realize economies of scope in goal achievement (Schader, 2009). Clearly, abolishing distorting subsidies is also a necessity, such as on mineral fertilizers. Much research has been done on such instruments in conventional contexts, but more research is needed for the optimal implementation of these instruments in specific contexts in organic agriculture.

Trade and sale

Just as more research on producers' acceptance of new technologies is needed, there is a need for research on consumers' acceptance of new technologies and, in particular, of new varieties of crops. This refers to resistant varieties in particular, that have the potential to offer solutions to certain pests and diseases, but that are not known to consumers and may taste different from well-known varieties. This is very important, as demand for the crops produced has to be assured. This will be less an issue with staple foods, where the primary goal is to provide food security. But it is of importance for fruits and vegetables and cash crops for export markets in particular. Part of this is the involvement of consumers, which is currently very minimal, especially in determining which varieties to promote. For both these cases regarding consumers' and producers' acceptance of technologies and varieties, participatory breeding methods would be recommended to improve the situation.

Research is also needed on how to optimally embed organic value chains in existing transport and trade infrastructure and logistics, which are designed for conventional production with much higher volumes and less demand for traceability of single products.

Power relations and inequalities

Finally, we mention *power relations and inequalities*. Research is needed on how to optimally design organic value chains, standards and certification to reduce inequalities and power asymmetries to a minimum. Some illustrations of such inequalities are given in Chapter 3, where certification can lead to exclusion of some farmers, for example. Another issue is monopoly power of certain players along the value chain, e.g. if conventional cotton can be ginned in five different ginneries, but only one is able to guarantee separated good flows and storage between organic and conventional cotton and thus becomes a monopoly. Currently, this is not a key topic in organic research. It could – and should – be an issue in the planned ORCA Centre on Economics, Markets and Trade, for example, but from the available description on this, it is not planned as a core topic. On inequality and power, we also reiterate the issues from UNEP-UNCTAD (2006), namely that more research is needed on how organic agriculture may benefit resource-poor households, and women and children in particular, and whether commercialization along organic value chains really has the potential to lead such households out of poverty, or whether and where a focus on subsistence production may be advantageous.

References

Altieri, M. (2005) 'The Myth of Coexistence: Why Transgenic Crops Are Not Compatible With Agroecologically Based Systems of Production', *Bulletin of Science, Technology and Society*, 25, 4, pp. 361–371

ASARECA (2009) *Natural Resource Management and Biodiversity Programme Vision and Strategy 2009–2014,* Association for Strengthening Agricultural Research in Eastern and Central Africa, Entebbe, Uganda

Bunders, J., Haverkot, B. and Hiemstra, W. (eds) (2005) 'Biotechnology: Building on Farmer's Knowledge', Macmillan Education, London

FAO (2008) 'Climate Change Could Increase the Number of Hungry People', Climate Change Report, www.fao.org/newsroom/en/news/2005/102623/index.html

Fuglie, K., Zhang, L., Salazar, L. and Walker, T. (1998) 'Economic Impact of Virus-free Sweet Potato Seed in Shandong Province, China', International Potato Center, Lima, Peru, www.eseap.cipotato.org/MF-ESEAP/Fl-Library/Eco-Imp-SP.pdf

Garrity, D.P., Akinnifesi, F.K., Ajayi, O.C., Weldesemayat, S.G., Mowo, J.G., Kalinganire, A., et al. (2010) 'Evergreen Agriculture: A Robust Approach to Sustainable Food Security in Africa', *Food Security,* 2, pp. 197–214

Godfray, H., Beddington, J., Crute, I., Haddad, L., Lawrence, D., Muir, J., et al. (2010) 'Food Security: The Challenge of Feeding 9 Billion People', *Science,* 327, 6

Hartmann, H.T., Kester, D.E. and Davies, F.T. (1990) 'Plant Propagation: Principles and Practices', Prentice Hall, London

Høgh-Jensen, H. (2011) 'Innovation research in value chains', *ICROFS News,* 2, pp 10–13

IAASTD (2008) 'Agriculture at Cross Roads – Global Report' www.agassessment.org/reports/IAASTD/EN/Agriculture at Crossroads_Global Report pdf

IBS (2005) 'IFOAM Basic Standards for organic production and processing, Version 2005', International Federation of Organic Agriculture Movements, http://www.ifoam.org/about_ifoam/standards/norms/norm_documents_library/IBS_V3_20070817.pdf

IFOAM (2002) 'Position on Genetic Engineering and Genetically Modified Organisms', International Federation of Organic Agriculture Movements, www.ifoam.org/press/positions/ge-position.html

IFOAM (2009) 'Organic Agriculture and Women's Empowerment', Cathy Farnworth and Jessica Hutchings, International Federation of Organic Agriculture Movements

Markwei, C., Ndlovu, L. and Robinson, E. (2008) 'International Assessment of Agricultural Knowledge, Science and Technology for Development', IAASTD, Global Summary for Decision-makers, p. 21

Nellemann, C., MacDevette, M., Manders, T., Eickhout, B., Svihus, B., Prins, A.G., and Kaltenborn, B.P., (eds), (2009) 'The Environmental Food Crisis – The Environment's Role in Averting Future Food Crises', A UNEP Rapid Response Assessment. United Nations Environment Programme, GRID-Arendal, www.grida.no

NOGAMU (2006) 'Research Needs Assessment for National Organic Agriculture Movement of Uganda NOGAMU', Report, August, Kampala, Uganda

Ogero, K., Gitonga, N., Ombori, O. and Ngugi, M. (2010) 'Cassava Production and Limitation of Propagation through Tissue Culture', in M. Mwangi (ed.) *Contributions of Agricultural Sciences towards Achieving the Millenium Development Goals,* FaCT Publishing, Nairobi, Kenya, pp. 148–155

ORCA (2009) 'Project Proposal – Start-up phase for the Organic Research Centres Alliance ORCA and Prototype Centre for Humid and Sub-Humid Areas (HUSHA)', ORCA, Rome, Italy

Paul, H. and Steinbrecher, R. (2003) *Hungry Corporations: Transnational Biotech Companies Colonise the Food Chain,* Zed Books, London

Pretty, J., Sutherland, W.J., Ashby, J., Auburn, J., Baulcombe, D., Bell, M., et al. (2010) 'The Top 100 Questions of Importance to the Future of Global Agriculture', *International Journal of Agricultural Sustainability*, 8, 4, pp. 219–236

Roeder, P. and Rich, K. (2009) 'The Global Effort to Eradicate Rinderpest', IFPRI Discussion Paper 00923, International Food Policy Research Institute (IFPRI), Washington DC, USA

Schader, C. (2009) 'Cost-effectiveness of Organic Farming for Achieving Environmental Policy Targets in Switzerland', PhD thesis, Institute of Biological, Environmental and Rural Sciences, Aberystwyth, Aberystwyth University, Wales. Research Institute of Organic Farming (FiBL), Frick, Switzerland, www.fibl-shop.org/shop/pdf/1539-cost-effectiveness-of-organic-farming.pdf

Ssekyewa, C. (2008) 'Challenges of Soil Fertility Building under the Organic Farming System in Africa', *Journal of Science and Sustainable Development*, 1, pp. 3–7

Ssekyewa, C. and Muwanga, M. (2009) 'Biotechnology in Organic Agriculture in Africa: Myth or Oversight?', *Journal of Science and Sustainable Development*, 2, 1, pp. 33–38

Ssekyewa, C., Opio, F.A., Swinburne, T.R., Van Damme, P.L. and Abubakar, Z.M. (1999) 'Sustainable Management of Collar Rot Disease of Passion Fruits in Uganda', *International Journal of Pest Management*, 45, 3, pp. 173–177

Stoll, G. (1998) *Natural Crop Protection in the Tropics*, AGRECOL, Margraf Verlag, Weikershein, Germany

TP-Organics (2009) 'Strategic Research Agenda for Organic Food and Farming', Technology Platform 'Organics'

TP-Organics (2011) 'Priority Proposals and Comments of TP Organics for the Forthcoming 2013 Calls of the EU 7th Research Framework Programme', Technology Platform 'Organics', Brussels, Belgium

Twarog, S. and Kapoor, P. (2004) 'Protecting and Promoting Traditional Knowledge: Systems, National Experiences and International Dimensions', United Nations, Geneva

UNEP-UNCTAD (2006) 'Overview of the Current State of Organic Agriculture in Kenya, Uganda and the United Republic of Tanzania and the Opportunities for Regional Harmonization', UNEP-UNCTAD Capacity Building Task Force on Trade Environment and Development (CBTF), United Nations

WDR (2008) *Agriculture for Development, The World Development Report 2008*, World Bank, Washington, DC

Witcombe, J.R., Virk, D. and Farrington, J. (eds) (1998) *Seeds of Choice: Making the Most of New Varieties for Small Farmers*, pp. 53–58, ITDG Publishing, Rugby, UK

Index